Human Factors Interventions for the Health Care of Older Adults

Human Factors Interventions for the Health Care of Older Adults

Edited by

Wendy A. Rogers
Arthur D. Fisk
Georgia Institute of Technology

LAWRENCE ERLBAUM ASSOCIATES PUBLISHING

2002   Mahwah, New Jersey   London

Lawrence Erlbaum Associates, Inc., Publishers
10 Industrial Avenue
Mahwah, NJ 07430

Cover design by Kathryn Houghtaling Lacey

Library of Congress Cataloging-in-Publication Data

Human factors interventions for the health care of older adults / edited by Wendy A. Rogers and Arthur D. Fisk.
    p. cm.
    Includes bibliographical references and index.
ISBN 0-8058-2300-X (alk. paper)
1. Aged—Medical care—Congresses. 2. Human engineering—Congresses. 3. Cognition in old age—Congresses. 4. Communication devices for the disabled—Congresses. I. Rogers, Wendy A. II. Fisk, Arthur D.

RC952.5 .H85 2001
362.1'9897—dc21   2001018977
                                          CIP

Books published by Lawrence Erlbaum Associates are printed on acid-free paper, and their bindings are chosen for strength and durability.

Printed in the United States of America
10 9 8 7 6 5 4 3 2 1

Dedicated to the many older adults who have given their valuable time to assist us in our research efforts. Together, we strive to improve the quality of life for all people.

Contents

Part III: Home Health Care and Caregiving

Part IV: Technology and Medicine

Part V: Computers and Health Care

Preface

In February, 2000, the Conference on Human Factors and Health Care Interventions for Older Adults was sponsored by the Center on Aging and Cognition: Health, Education, and Training (one of the Edward R. Roybal Canters for Research on Applied Gerontology funded by the National Institute on Aging). In addition to the speakers, other members of the research community participated in the discussion periods. The conference attendees are pictured on p. x. This volume represents the culmination of the discussions that took place at this conference.

The unifying theme of this volume is the relevance and contributions made by the field of human factors to health care interventions. The opening chapter provides an overview of the critical issues and the needs for human factors research. The second chapter provides an overview of the techniques, tools, and methods of the field. The next three chapters present the cognitive issues that must be considered in the context of health care environments such as memory abilities, linguistic process, and decision making. The following chapter provides an illustration of how cognitive processes showing age-related declines may be enhanced through an exercise program. The remaining chapters cover a wide range of cutting-edge topics including caregiving, telecommunications issues, design of medical devices, computer monitoring of patients, automated communication systems, computer interface issues in general, and the use of the World Wide Web as a source for health information.

The conference and the book were the products of the time and energy expended by many individuals. First, we would like to thank the presenters as well as the other participants in the conference for helping us to make the conference a success. We also acknowledge the many individuals who assisted in various aspects of the conference. Denise Park, Denise Taylor-Moon, and Janet Peake all were very helpful throughout the planning of the conference. Also, we thank the students who helped us before, during, and after the conference: Holly Hancock, Tim Nichols, Rich Pak, and Aideen Stronge.

The editorial staff at Lawrence Erlbaum Associates has been very supportive and professional, as always. We especially thank Anne Duffy and Marianna Vertullo. We would also like to thank the following individuals for taking the time to assist in the review process: Barry Beith, Jim Callan, Darin Ellis, Don Fisher, Darlene Howard, Kathie Insel, Alex Kirlik, Don Lassiter, Rob Mahan, Sherry Mead, Beth Meyer, Tom Pierce, Mark Scerbo, Liz Stine-Morrow, and Gail Williamson.

Finally, we thank each of the contributors for their dedication to the goal of improving the health care of older individuals.

—*Wendy A. Rogers*
—*Arthur D. Fisk*

Center for Aging and Cognition: Health, Education, and Training Conference on Human Factors Interventions for the Health Care of Older Adults, February 24–26, 2000, Destin, FL.

Pictured left to right:

First Row: Dan Fisk, Wendy Rogers

Second Row: Peter Hancock, Aideen Stronge, Odette Gould, Sue Bogner, Constance Qualls, Holly Hancock, Richard Pak, Joyce Harris

Third Row: Tom Sheridan, Art Kramer, Daryle Gardner-Bonneau, Neff Walker, Max Vercruyssen, Sara Czaja, Kathleen Insel, Beth Meyer, Patricia Holley

Fourth Row: Andy Smith, Richard Schulz, Alex Chaparro, Scott Brown, Tim Nichols, Eric Stephens, Don Lassiter, Barry Beith, Dan Morrow, Neil Charness, Jim Fozard, Pam Whitten, Alex Kirlik

Not pictured: Ann Benbow, Russell Ellison, Denise Park, Janet Peake, and Denise Taylor-Moon.

PART I

OVERVIEW AND GENERAL ISSUES

1

Health Care of Older Adults: The Promise of Human Factors Research

Arthur D. Fisk

Wendy A. Rogers

Georgia Institute of Technology

It has been made clear in the scientific and popular press that the average age of the population in developed countries is increasing. The number of older adults is increasing faster than that of their younger counterparts. Rowe and Kahn (1998) emphasized the dramatic increase in life expectancy by pointing to an estimate showing that of all the humans who have ever lived to the age of 65 years or older, half are currently alive today. Such a change in demographics brings with it unique challenges and opportunities for both the private and public sectors (Fisk, 1999). Providing affordable, safe, and usable health care systems, especially for home-based usage, is one such challenge and opportunity. The chapters in this volume collectively point the way for building on the established foundation and meeting those challenges.

Businesses must deal with the shifts in demands for goods and services brought on by the desires and needs of older consumers (Howell, 1997). The issues facing the delivery of goods and services in general are magnified in the health care arena. There are several areas of concern. Only a few are mentioned in this discussion. Efforts to reach older consumers often are hampered by failure to understand the requirements and preferences of this population. For example, in the past, automatic teller machines were designed with little regard for older users (Rogers & Fisk, 1997). The consequences of poor design for banking machines may be inconvenience, but when it comes to health care products, the consequences can be life threatening (Bogner, 1994).

The lack of solid usability analyses for the physical design of systems cascades into the design of tasks and training. For example, systems as relatively constrained as

1

blood glucose monitors often are advertised "as easy to use as 1, 2, 3," but task analysis indicates that the three steps explode to at least 50 steps (Rogers, Mykityshyn, Campbell, & Fisk, in press). Furthermore, an accurate reading depends on error-free performance of each step. In an empirical evaluation of blood glucose monitor use, older adults (naïve to the use of such systems), even with training, averaged only 70% correct when required to do no more than calibrate the device (Mykityshyn, 2000). Gardner-Bonneau and Gosbee (1997) succinctly illustrated the point that an alarming number of devices meant to provide home medical care are prone to serious error by older adults. The challenge is to design products and services so that users can realize maximal performance potential and be assisted in doing so by the design, rather than hindered by design induced errors. Proper human factors research conducted to advance both theory and practice can bring this to fruition (Fisk & Kirlik, 1996).

THE PROMISE OF HUMAN FACTORS

Historically, the design of technology has been the responsibility of engineers, and the discovery of fundamental human capabilities and limitations has been a task for the behavioral scientist. The demands of designing technology for older adults, in terms of both opportunities and risks, clearly point the way for the behavioral scientist to take the lead more actively in developing design specifications for everyday and advanced technology. To design effective systems and training programs in support of age-related performance capabilities, fundamental questions must be answered concerning aging and complex task performance.

The past several decades have produced much information on the process of becoming an older adult. Clearly, much more is known about aging than is being usefully applied, but it is equally clear that huge gaps exist in our knowledge. One might ask what human factors have to do with a focus on older adults. The discipline of human factors can be defined as the development and application of human–system interface technology (Hendrick, 1997). Human–system interface technology deals with the interfaces, and training in the use of those interfaces, between humans and other system components including complex hardware, computer software, environments, jobs, pill bottles, instructions, warnings, and so on. As a science, human factors studies human capabilities, limitations, and other characteristics for the purpose of developing human–system interface and training technology. As a practice, human factors applies human–system interface technology to the analysis, design, evaluation, standardization, and control of systems. The development of training technologies is part and parcel of the human factors technology (for a historical review see Flanagan, 1948; Wolfle, 1951). Instructional design has been proved effective for enhancing performance with even the optimum interface design (see Fitts & Posner, 1967 for a review). Human factors, as part of its history, used its unique design and training technology to ensure that pilots could safely and effectively perform in the cockpit of high-performance jet fighters, that complex systems such as nuclear power plants could be controlled more effectively, and so on. Certainly, this unique technology can be used to guide the design of systems to accommodate the aging user better within the health care arena.

Often, the aging process simply exacerbates suboptimal design features that, to a lesser extent, affect everyone's learning, performance, and well-being. The basic premise connecting human factors with information on aging is that the world in which people live, work, play, and get around must be designed with systematic regard for the older user. As people age, they are exposed to an increasing array of threats to their health, safety, performance, and quality of life, many of which could be avoided or minimized through improved design. Improvement is achieved by translating fundamental knowledge of age-related changes in human capabilities, tendencies, and preferences into design principles or requirements. To use a straightforward example, Walker and his colleagues (Walker, Philbin, & Fisk, 1997; Walker, Spruell, & Philbin, 1996; Worden, Walker, Bharat, & Hudson, 1997), demonstrated how understanding of age-related changes in human movement control could be applied to the design of an adaptive interface to allow easy and efficient computer input control across the adult life span.

As mentioned earlier, a critical aspect of the human factors discipline is the development of instruction and training. Although system designs can almost always be improved, even the most well-designed system (e.g., automobiles and computers) will likely require some training for novice users. Furthermore, training often is required for optimum use of even well-designed systems (again, consider automobiles and computers). The aging process also may result in the need for age-specific instructional designs that enable older individuals to interact safely and effectively with systems at home and at work.

Design and training interventions based on human factors methods and principles have the potential to solve many problems that older adults experience in their daily activities. Rogers, Meyer, Walker, and Fisk (1998) conducted a series of focus groups wherein they asked adults between the ages of 65 and 88 years to discuss the types of difficulties they encounter in their everyday activities. These were community-dwelling individuals who lived in single-family houses, apartments, condominiums, or retirement communities (e.g., a high-rise). The older adults reported a number of constraints that limited their activities, which Rogers et al. (1998) categorized as follows:

- *Motor limitations*: difficulty getting on a bus, walking, bending, climbing stairs, and so on

- *Perceptual limitations*: vision changes influencing reading or driving, hearing loss, and so on

- *Cognitive limitations*: difficulty understanding how to operate a device, forgetting information, confusion about options in telephone menus, and so on

- *External limitations*: fear of crime, financial limitations, family responsibilities, and so on

- *General health limitations*: health-related declines in functional activity.

Rogers et al. (1998) also classified each problem reported by their participants in terms of whether the issue could potentially be addressed by human factors professionals. Those data are summarized in Table 1.1. Approximately 7% of the reported problems could be remedied through some type of training intervention: These problems typically were related to cognitive issues such as unfamiliarity with a particular task or system. Design interventions could possibly remedy 25% of the problems, primarily those related to sensory or motor difficulties.

One striking outcome of the Rogers et al. (1998) data was the frequency with which older adults reportedly came into contact with new technologies. The problems reported with such technologies could be overcome most effectively using interventions that improved the design of the system itself to enhance usability, incorporating a training intervention to provide instruction for using the system (21% of the problems documented in the Rogers et al. [1998] study fell into this category). Many complex systems such as computers could benefit from redesign, but also will likely require training as well for the novice user. Thus, approximately 53% of the reported problems encountered by older individuals could be minimized potentially through human factors interventions. The types of problems found to be less amenable to such interventions involved issues related to loss of a close family member, financial difficulties, and so on.

The Rogers et al. (1998) results were not meant to imply that human factors professionals have all the answers ready at hand to solve every problem that older adults might have. However, the field clearly has the potential to develop such solutions. In fact, in the health care arena, the chapters in this volume illustrate some of the solutions, along with directions for the development of future solutions.

The underlying goal, we believe, of human factors intervention, especially concerning delivery of health care, was summed up eloquently by Rabbitt (1992): "Demographic changes make it vital for designers to become aware of the nature and extent of age changes in physical, sensory, and cognitive abilities. The fact that these changes are complex and interact with each other in subtle ways make their study intellectually fascinating as well as humanely useful. The goal of helping each other to enjoy independence during the extra years that medical and social advances have won for us is surely as rewarding as concentrating on increasing sales and market penetration. Further, ... these goals are entirely compatible" (p. 137). Rabbitt (1992) nicely pointed to the fact that the health, safety, and well-being of older adults are a fundamental concern of human factors. Business practice also is enhanced. However, this is a by-product of the main goal.

TRAINING INTERVENTIONS

There is a myriad of technologies, both new and established, with which older individuals interact in the context of caring for themselves or others. What is the best way to train older individuals to use these systems? Usually, it is not sufficient to design a training program for young adults and use it for older adults as well. Evidence suggests that training programs may be differentially effective for different age groups (Jamieson & Rogers, 2000; Mead & Fisk, 1998).

TABLE 1.1

Potential for Human Factors Interventions

Intervention	Primary Issues	Examples	Problems Remediable (%)
Training alone	Cognitive	Learning to drive, way finding, doing medical procedures, learning to use public transportation	7
Redesign alone	Sensory or motor	Steps on bus, need for tools for grasping or scrubbing, opening medical bottles, seeing displays	25
Both training and redesign	Complex	Exercise equipment, hope electronics, kitchen appliances, insurance and banking forms	21
			Total = 53

An understanding of age-related changes in cognition is important for the development of age-appropriate training programs. Based on his review of age-related changes in memory, Smith (chap. 3, this volume) was able to make the following recommendations relevant, particularly, to the development of training for older adults:

- Take advantage of increased world knowledge, expertise, and semantic memory.

- Use well-integrated and familiar cues.

- Avoid memory demands, where possible, through provision of cues.

- Provide more extensive training for older adults.

- Minimize irrelevant or distracting information from instructions.

Linguistic abilities also must be considered in the development of training and instructional materials. Both written and spoken language comprehension are necessary for health care tasks such as following instructions from a nurse or physician or learning how to use a new medical device or fill out insurance forms. Qualls, Harris, and Rogers (chap. 4, this volume) provide guidelines for minimizing the influence of age-associated comprehension declines. For instance, providing contextual cues, using simple direct sentences, and even ensuring that fonts are large enough and sounds are clear enough all can improve the comprehension of older readers and listeners. In addition, older individuals may be taught strategies that they may implement themselves when trying to interpret instructional materials or comprehend complex information. The Qualls et al. report documents evidence of the benefits from rereading, either selectively or in full, and from the use of elaborative, memory-enhancing strategies.

Accurately assessing the benefits of training is necessary for making selections among training options. Charness and Holley (chap. 13, this volume) describe one method for assessing the effectiveness of training through the use of learning curve data. Predictions can be made about the number of trials that will be necessary to reach a particular level of performance by extrapolating from a learning curve based on measured rates of learning for a particular group (e.g., older adults). Rates of learning can be compared statistically to make estimations about the benefits from different types of training. In addition, estimates can be made about how much training will be needed for a desired level of performance. Such predictions may be very useful in illustrating the cost effectiveness of training programs.

The discussion so far has focused on how older individuals can best be trained to use systems. However, another form of training intervention can be conceptualized as training older adults to "develop themselves" through an exercise program as a method to maximize their health and cognitive functioning. This was the general approach taken by Kramer and his colleagues (chap. 6, this volume). They investigated the cognitive benefits of engaging in an exercise program. Their primary question was whether exercise would improve cognitive function in older adults who previously were sedentary. Importantly, their results showed that the type of exercise was a critical variable. A 6-month exercise intervention was most beneficial if the exercise program was aerobic (walking) rather than nonaerobic (stretching and toning). Moreover, not all cognitive functions showed improvement. The effects were most evident for cognitive functions that required executive function (e.g., planning, monitoring) and presumably relied on the frontal and prefrontal cortex. As Kramer et al. state, a logical next step for this research will be to determine the degree to which aerobic exercise benefits daily activities such as monitoring a medication regimen or caring for a spouse.

DESIGN INTERVENTIONS

General Intervention Issues

Design interventions based on a fairly broad use of the term "design" are discussed. That is, the design of specific products or devices is included, as well as the design of systems such as telemedicine or caregiver aid programs.

To begin with, designers should capitalize on the knowledge base underlying the field of cognitive aging. There is a solid foundation of data about the perceptual, motor, and cognitive changes that occur as individuals age (for recent reviews, see Craik & Salthouse, 2000; Fisk & Rogers, 1997; Park & Schwartz, 2000).

Charness and Holley (chap. 13, this volume) describe the perceptual and motor issues that should be considered for computer interfaces intended for use by older adults. Visual declines are common in older adults, and given that much of the information in computer displays is presented visually, such declines must be considered in the design process. There also are auditory declines and disorders that occur more frequently for older adults that may influence speech comprehension and the ability to process other auditory information such as alarms. Motor changes include slowing of response speed and declines in coordination and dexterity. These changes have direct bearing on the ability of older individuals to control input devices on computers or to manipulate small components of medical devices.

Designers also need to consider age-related cognitive limitations. For example, it is well known that older adults have memory problems when required to recall information without the benefit of cues. Smith (chap. 3, this volume) describes the importance of providing, within the design of the system, "environmental support" to minimize the cognitive demands of the user. One simple example of an environmental support is the provision of memory cues at critical times in a sequential task when a particular action must be carried out.

Another example of environmental support is provided by Stronge, Walker, and Rogers (chap. 14, this volume) in the context of Web design. They discuss the potential benefits of providing a support for search tools such that users need not worry about the specific format of their search, but instead are prompted for options such as "and" or "or" to expand or constrain a search for information.

Designers must be aware of the user's linguistic capabilities and make efforts to accommodate age-related changes in written and spoken language comprehension (Qualls et al., chap. 4, this volume). Some of the difficulties that older users experience will result from perceptual declines in vision and hearing. These too must be accommodated. Written information should be designed and tested to ensure readability by older users, with consideration given to such variables as font size and type, contrast, use of color, and issues of glare. Spoken information must be organized in line with the expectations of the listener; sentences should be short and presented in active voice; and so on. Morrow and Leirer (chap. 10, this volume) have implemented these guidelines in their system with much success.

It is important to recognize that not all abilities decline with age. In fact, aspects of memory and linguistic abilities are maintained or even improved as individuals grow older. Abilities that tend to remain intact into old age include some aspects of memory (e.g., recalling well-learned information), verbal abilities such as vocabulary and reading, and some aspects of attention (e.g., focusing on a single source of information). Part and parcel of the definition of human factors, understanding these abilities and capitalizing on them can be as important as compensating for declining abilities.

Specific Intervention Issues

The term "telemedicine" or "telehealth" refers to the broad area of providing health care across distances. The discussions in this volume focus on the circumstances in which either the caregiver or the recipient of the care is an older adult. Whitten (chap. 7, this volume) provides an overview showing the general characteristics of such care, the potential benefits for older adults, and the specific concerns of older adult users of the system. However, her review of the literature to date makes it clear that few empirical studies have been conducted to document the specific benefits of telemedicine, how systems need to be redesigned to accommodate the variety of potential users, or even whether telemedicine provides a viable, cost-effective alternative to traditional methods of health care.

Whitten is hopeful that current and future studies will provide insight into the many questions surrounding the use and benefit of telemedicine generally, as well as specifically for older adults. She provides questions that such studies should specifically address, only a few of which are mentioned:

- How will new systems be integrated with existing health care products and services?

- What design elements enhance credibility and ease of use?

- How are issues of security and confidentiality to be resolved?

- Will health outcomes be impacted as clinical care and services are made more available via the Internet?

Some of these questions are being answered already in the context of designing telecommunication systems to provide health care to older adults. One area in which this research is being conducted is family caregiving: providing assistance to individuals debilitated in some way as a result of disease or accident. Schulz, Czaja, and Belle (chap. 8, this volume) illustrate the complex array of psychosocial issues that can influence the success or failure of interventions. Relevant variables include the characteristics of the caregiver (e.g., age, gender, ethnicity, experience), the types of caregiving tasks required (e.g., bathing, performing medical procedures), and the resources available (e.g., financial independence, support from other family members, supplemental nursing care). Schulz et al. have developed a model that enables prediction about who will experience caregiver stress, when it will be experienced, and what type of intervention will be most beneficial. On the basis of this model, they characterize potential interventions and provide a method for measuring the benefits of interventions.

The benefit of using information technology as an aid to caregiving is discussed by Czaja (chap. 9, this volume). She reports initial findings from the Resources for Enhancing Alzheimer's Caregiver Health (REACH) program, which is evaluating the efficacy of various caregiver interventions. The Miami site is evaluating the use of a computer–telephone system for family caregivers. The system is a hierarchically structured, menu based telephone system that enables caregivers to communicate with family, friends, therapists, and an online support group, and to access services. The goals of the REACH project are to assess the specific benefits of the technology, to evaluate how it is integrated into the health care systems, and to conduct a usability evaluation of the technology itself. On the basis of the initial findings from the study, Czaja is able to make system design recommendations for future versions of the system. This type of project represents the instantiation of a technology-based intervention conducted and evaluated by human factors experts.

The potential for telephone systems to augment health care communication also is being evaluated by Morrow and Leirer (chap. 10, this volume). Their system is designed to aid patients in receiving communications from their physicians that relate to their appointments such as activities to be done before the appointment, details of the appointment, and follow-up information. Morrow and Leirer have attempted to apply their understanding of age-related changes in cognition (e.g., attention, comprehension, memory limitations) to the development of an easy-to-use automated system. Their work is an example of a user-centered design that incorporates basic principles of human factors such as needs analysis and usability testing. The development of their system has progressed through an iterative process whereby initial versions were tested and subsequently modified on the basis of user experiences. They also report the success of their system in influencing the behavior of the users by improving appointment attendance and preparedness. Their work provides existing proof of the utility of technology for benefiting the health care of older individuals. Moreover, their design strategy may be generalizeable to other systems that have similar goals and functions for an older adult user population.

Clearly, the trend for telemedicine to augment or replace traditional methods of health care necessitates the development of technological systems that will accomplish various tasks such as providing caregiver support or physician appointment information. Monitoring patients is another function that may be carried out long-distance via computer technology. The potential for sensor and computer technology to monitor patients with congestive heart failure is being evaluated by Sheridan, Coughlin, Kim, and Thompson (chap. 11, this volume). Particular strengths of their monitoring device are that it can be calibrated to individual patients, measure whether their condition is stable or deteriorating, and automatically send an alert to the hospital. Not surprisingly, then, a device of this complexity is not trivial to develop.

Sheridan et al. have developed detailed decision logic for their diagnostic system and use mathematical modeling techniques to determine changes in the physiologic state of the patient. They also have considered usability issues from the patient's and physician's perspectives in the development of their respective displays, tried to minimize the need for training, and designed the system to self-diagnose when it is not working properly or is getting incorrect information. As with other design approaches reported in this volume, the Sheridan et al. work follows the tenets of human factors and has the potential to provide guidelines for the development of future systems.

Usability considerations for medical devices begin with their initial setup and periodic adjustment or calibration procedures, involve the proper use of the device for its intended purpose, and extend to the cleaning and maintenance of the device over time (Hyman, 1994). Design decisions should be made to minimize opportunities for errors for all interactions with the device, from beginning to end, with consideration given to who will be using the device and where they will be using it. Gardner-Bonneau (chap. 12, this volume) describes the human factors issues surrounding the development, testing, and use of medical devices. Her chapter provides insight into the industry and the government regulating bodies. It is encouraging that the Food and Drug Administration now requires that device designs must consider the needs and capabilities of the patient, in essence the application of human factors principles to design.

Research reviewed to date by Gardner-Bonneau has documented many human factors problems that exist in current medical devices. These issues provide a starting point for the improvement of future devices. Most important, perhaps, is to ensure that a human factors professional is included on the design team and that appropriate usability testing is carried out. However, whereas the need for human factors input into device design is being recognized by the industry and by the government, the mechanisms for providing this input are in the nascent stages. Gardner-Bonneau reviews recent activity in the development of standards that will guide medical device design and the development of a handbook that will transmit human factors design information to the manufacturers of the devices.

The development of standards and guidelines should perhaps be pursued also for the World Wide Web. The Web is becoming a source of health-related information, credible or not, for individuals of all ages. Unfortunately, the design of Web sites and search tools has focused little on issues of usability. The usability problems most relevant to older users are reviewed by Stronge, Walker, and Rogers (chap. 14, this volume). Many older adults already use or would like to use the Web, and the acquisition of health-related information is an oft-cited reason (Morrell, Mayhorn, & Bennett, 2000). However, Stronge et al. suggest that there is a relatively low probability that the average older adult would be successful in finding specific health information on the Web. The complexity of the system, the sheer amount of information available, and the lack of user-centered design all contribute to the problem. Stronge et al. recommend capitalizing on knowledge of cognitive aging, available design guidelines, and usability testing of the targeted user populations as a starting point for improving the usability of the web for older adults. The Web has much potential to provide health care information, and even health care services, but human factors issues must be addressed before this potential will be realized.

REFINING HUMAN FACTORS METHODS

As demonstrated throughout the chapters in this volume, human factors techniques and tools such as needs analysis, usability testing, instructional design methods, and so on have much promise for addressing delivery of health care to older adults. At the same time, one avenue for future progress is the development of new methods to be developed and the refinement of existing methods. Beith (chap. 2, this volume) emphasizes the need for innovation in efforts to understand the needs of older adults in the health care arena and to accommodate those needs. He recommends techniques such as immersive experiential, ethnography, and video data analysis in conjunction with understanding the technological advances defining the user behaviors intended to be aided by such technologies.

A general method for approaching, understanding, and solving problems experienced by older adults in the context of health care is advocated by Kirlik and Strauss (chap. 5, this volume). They describe an ecological approach wherein a systems view is used to understand difficulties experienced by older individuals in the context of performing health care tasks (e.g., the decision-making processes involved in selecting a long-term care provider). In itself, the ecological approach is not new, although instantiation of it and the application to this problem domain by Kirlik and Strauss is new and demonstrates the theoretical and practical power of their paradigm. Many factors influence decisions. The individual and interactive effects of these variables must be considered for interventions to be successful. The ecological perspective enables specific (i.e., quantitative) predictions about when errors will occur, and hence provides direction for where interventions should be targeted.

CONCLUSION

The National Research Council recently published a report entitled, "The Aging Mind," designed to make recommendations for future cognitive research (Stern & Carstensen, 2000). One of their recommendations is for "developing the knowledge needed to design effective technologies to support adaptivity in older adults" (p. 35). This laudable goal has direct relevance to the issues of improving the health care of older individuals. The report also emphasizes that for such a goal to be realized, "it requires integrating behavioral science and engineering in a context of product design and development" (p. 36). This statement bears striking resemblance to a subset of professional activities in the field of human factors, activities carried out by human factors specialists on a daily basis for the design of systems with a range of diversity from computers to farm tractors. Clearly, as demonstrated in the chapters of this volume, human factors interventions can improve the health care of older individuals. It also seems clear from the individual chapters that although much remains to be accomplished, both the science and practice of human factors hold much promise to meet these important challenges.

ACKNOWLEDGMENTS

The authors were supported in part by grants from the National Institutes of Health (National Institute on Aging): Grant P50 AG11715 under the auspices of the Center for Aging and Cognition: Health, Education, and Technology, Grant P01 AG17211 under the auspices of the Center for Research and Education on Aging and Technology Enhancement (Edward R. Roybal Centers for Research on Applied Gerontology); Grant R01 AG07654, and Grant R01 AG18177. The authors are grateful to Barry Beith for constructive comments on the chapter.

REFERENCES

Bogner, M. S. (1994). *Human error in medicine*. Hillsdale, NJ: Lawrence Erlbaum Associates.
Craik, F. I. M., & Salthouse, T. A. (2000). *The handbook of aging and cognition* (2nd ed.). Mahwah, NJ: Lawrence Erlbaum Associates.
Fisk, A. D. (1999). Human factors and the older adult. *Ergonomics in Design, 7*(1), 8–13.
Fisk, A. D., & Kirlik, A. (1996). Practical relevance and age-related research: Can theory advance without practice? In W. A. Rogers, A. D. Fisk, & N. Walker (Eds.), *Aging and skilled performance: Advances in theory and application* (pp. 1–15). Mahwah, NJ: Lawrence Erlbaum Associates.
Fisk, A. D., & Rogers, W. A. (1997). *Handbook of human factors and the older adult*. San Diego, CA: Academic Press.

Fitts, P. M., & Posner, M. I. (1967). *Human performance*. Belmont, CA: Brooks/Cole.

Flanagan, J. C. (1948). *The aviation psychology program in the Army Air Force*. Aviation Psychology Program Research Report 1. Washington, DC: U.S. Government Technical Printing Office.

Gardner-Bonneau, D., & Gosbee, J. (1997). Health care and rehabilitation. In A. D. Fisk & W. A. Rogers (Eds.), *Handbook of human factors and the older adult* (pp. 231–255). San Diego, CA: Academic Press.

Hendrick, H. W. (1997). *Good ergonomics is good economics*. Santa Monica, CA: Human Factors and Ergonomics Society.

Howell, W. C. (1997). Forward, perspectives, and prospectives. In A. D. Fisk & W. A. Rogers (Eds.), *Handbook of human factors and the older adult* (pp. 1–6). San Diego, CA: Academic Press.

Hyman, W. A. (1994). Errors in the use of medical equipment. In M. S. Bogner (Ed.), *Human error in medicine* (pp. 327–347). Hillsdale, NJ: Lawrence Erlbaum Associates.

Jamieson, B. A., & Rogers, W. A. (2000). Age-related effects of blocked and random practice schedules on learning a new technology. *Journals of Gerontology: Psychological Sciences, 55B*, p. 343–353.

Mead, S. E., & Fisk, A. D. (1998). Measuring skill acquisition and retention with an ATM simulator: The need for age-specific training. *Human Factors, 40*, 516–523.

Morrell, R. W., Mayhorn, C. B., & Bennett, J. (2000). A survey of World Wide Web use in middle-aged and older adults. *Human Factors, 42*, 175–182.

Mykityshyn, A. (2000). *Toward age-related training methodologies for sequence-based systems: An evaluation using a home medical device*. Unpublished Masters' Thesis. Atlanta, GA: Georgia Tech.

Park, D. C., & Schwarz, N. (2000). *Cognitive aging: A primer*. Philadelphia: Psychology Press.

Rabbitt, P. M. A. (1992). Cognitive changes with age must influence human factors design. In H. Bouma & J. Graafmans (Eds.), *Gerontechnology* (pp. 113–140). Amsterdam: IOS Press.

Rogers, W. A., & Fisk, A. D. (1997). Automatic teller machines: Design and training issues. *Ergonomics in Design, 5*, 4–9.

Rogers, W. A., Meyer, B., Walker, N., & Fisk, A. D. (1998). Functional limitations to daily living tasks in the aged: A focus group analysis. *Human Factors, 40*, 111–125.

Rogers, W. A., Mykityshyn, A. L., Campbell, R. H., & Fisk, A. D. (in press). "Only three Easy Steps?" User-centered analysis of a "simple" medical device. *Ergonomics in Design*.

Rowe, J. W., & Kahn, R. L. (1998). *Successful aging*. New York: Pantheon.

Stern, P. C., & Carstensen, L. L. (2000). *The aging mind: Opportunities in cognitive research*. Washington, DC: National Academy Press.

Walker, N., Philbin, D. A., & Fisk, A. D. (1997). Age-related differences in movement control: Adjusting submovement structure to optimize performance. *Journal of Gerontology: Psychological Sciences, 52B*, P40–P52.

Walker, N., Spruell, C., & Philbin, D. A. (1996). The use of signal detection theory in research on age-related differences in movement control. In W. A. Rogers, A. D. Fisk, & N. Walker (Eds.), *Aging and skilled performance: Advances in theory and application* (pp. 45–64). Mahwah, NJ: Lawrence Erlbaum Associates.

Wolfle, D. (1951). Training. In S. S. Stevens (Ed.), *Handbook of experimental psychology* (pp. 1267–1286). New York: Wiley.

Worden, A., Walker, N., Bharat, K., & Hudson, S. (1997). Making computers easier for older adults to use: Area cursors and sticky icons. In *Proceedings of the Human Factors in Computing Systems '97 meeting*. New York: ACM.

2

Needs and Requirements in Health Care for the Older Adult: Challenges and Opportunities for the New Millennium

Barry H. Beith
HumanCentric Technologies, Inc.

With the arrival of the new millennium comes a renewed interest and energy in addressing some of our most complex social and technological issues and solving some of our most perplexing and enduring system problems. One such issue is the provision of adequate health care for people, and of particular focus in this volume, the elderly. As suggested in the introductory chapter by Fisk and Rogers, the issue is made more acute by the realization that the elderly segment of the population is growing at a rapid pace. Howell (1997) pointed out that by 2030, 22% of the U.S. population (66 million people) is expected to be older than 65 years and that the age group older than 85 years represents the fastest growing subgroup in the country. The irony and dilemma is that the very medical science that is prolonging life so dramatically is also creating a tremendous burden on itself to provide medical products and services. It is a very real concern that the elderly population's needs and requirements could strain the present health care system to the breaking point.

The health care system comprises a myriad of complex tools, products, systems, processes, and information that can easily overwhelm the user, whether patient or professional. The system has been described by Dr. Lloyd Hey (personal communication, August 28, 2000) of the Duke University Medical Center and founder of MDeverywhere.com as "a trillion dollar cottage industry" with thousands of providers of products and services that do not communicate or coordinate very well. Few systems in our country are more complex than the health care system with its tens of thousands of products, procedures, tools, organizations, personnel, institutions, regulations, and layers. Despite tremendous advancements in medical research and treatments, it is this complexity that often is responsible for diminishing the quality of health care that reaches individuals. This problem is magnified for older adults who bring reduced cognitive processing capabilities and increased medical needs to the system simultaneously. In a system wherein professionals often are challenged to keep up with the information and technology, it is not reasonable to assume that the elderly can do so.

To improve the health care system, planners, designers, and developers must have a solid foundation of information about the needs of the elderly and the system requirements to meet those needs. We are witnessing an explosion in technologies associated with database management, wireless telecommunications, and information use that potentially provide the elements of an effective infrastructure capable of addressing these enduring system problems in new and vigorous ways.

Needs assessment and requirements analysis are the most important activities for initiating system improvement because, done well, they are the foundation on which all other activities build. The more complex the system, the more critical these activities are that provide the foundation for good design. Unfortunately, despite the importance of these activities, they often are the most overlooked and underresourced activity in product and system development.

WHAT ARE NEEDS AND REQUIREMENTS?

Needs and requirements often are used interchangeably in practice. This obscures the fact that they are different in nature and based on different types of information inputs and outputs. Each activity is "necessary but not sufficient" to provide the foundation for defining and designing a product or system. Needs assessment without requirements analysis provides goals of the product or system, but without an implementation plan. Requirements analysis, on the other hand, without needs assessment can implement a design of a product or system, but without direction. Clearly, these activities are closely interrelated and dependent on each other.

Needs assessment is an initial activity that focuses on collecting, aggregating, integrating, and prioritizing information defining the needs of the individuals for whom some design is intended, e.g., a health care system for the elderly. Needs represent both intermediate and ultimate outcomes desired by or necessary for the individual to accomplish a goal. For example, if an elderly patient wants to remain at home or in his or her children's home for care rather than go to a hospital, nursing home, or other institution for care, there are several needs that can be identified to achieve that end. These needs might include information needed by the caregivers or the patient, equipment or medicines needed to manage a condition, or training on how to provide the care needed. Needs assessment as an activity focuses on bringing together all sources of information that may contribute to a comprehensive picture of what the elderly or their caregivers might need relative to providing effective home health care. Ideally, the needs assessment is conducted to identify information needs and task needs independently of existing procedures, tools, processes, and other current approaches to addressing appointment-related needs. Identifying this need does not necessarily require identifying the method or process by which the need will be met.

Requirements analysis takes information from the needs assessment and other sources and distills it into a set of requirements for the product or system to be designed. Requirements represent some characteristic or attribute of the system that is necessary to meet identified needs. This could include the design of information, such as warnings or instructions, as well as tools and equipment, processes, tasks, facilities, or procedures. The requirements analysis provides the input for initial task specifications necessary to the design, which then are added to other types of specifications from other disciplines such as engineering or programming. Unlike the information from the needs assessment, the requirements analysis output feeds directly into a specification of the product or system. In other words, the requirements drive the specification that describes how the product or system will meet the needs identified earlier. It is this input that allows the engineers, programmers, and other development groups to "build" the product, process, procedures, information, or system. In the current example of home care, a requirement might be to provide a specific product for monitoring biomedical vital signs on a regular basis, performing dialysis in the home, or providing adequate communication and monitoring system capabilities between the caregivers and patient.

Methods of requirements analysis typically are "homegrown" in that companies or agencies define their own approach to accomplishing this task and determining how it relates to any needs assessment that might have been conducted. Guidance is available as to how requirements analysis is conducted (Institute of Electronic and Electrical Engineers, 1998; Martin & Bahill, 1996), but no standardized approaches that have been widely accepted.

One approach to requirements analysis publicized in the Harvard Business Review was entitled the "House of Quality" (Hauser & Clausing, 1988). This approach was developed by the Japanese to facilitate interdisciplinary team discussions for converting needs and desires into requirements and specifications. Multidisciplinary matrices were developed to associate specifications with requirements and to facilitate different development participants in prioritizing competing solutions. The approach allowed compromises to be made in the design before the final specification was generated and proceduralized the generation of requirements. This represents one of the few efforts to provide a systematic requirements and specifications tool and, although praised as effective, even the House of Quality does not appear to have provided a widely adopted standardized approach to requirements analysis.

TRADITIONAL APPROACHES TO DETERMINING NEEDS AND REQUIREMENTS

Traditional approaches to needs assessment have been used for decades and typically are very straightforward. These approaches include surveys, interviews, and focus groups.

Although it is not within the scope of this brief discussion to address the wide variety of approaches to conducting surveys, there have been several good sources for doing so over the years (Dillman, 1999; Meister, 1985; Rea, Parker, & Schrader, 1997;

Warwick & Lininger, 1975). The primary virtue of this technique is in the ability to collect information from large numbers of individuals in a relatively short period of time. Many difficulties must be overcome, however, regarding response rates, self-selection effects, various biases associated with poorly designed survey devices, and poorly understood survey items (Lee, Forthofer, & Lorimor, 1990). Designing good surveys is a craft in itself. Designing good surveys for an elderly population may be even more difficult because of the decline in this group's cognitive processing abilities. In the case of home health care for the elderly, a survey might be undertaken to identify the information needs of home care providers in providing health care to their elderly family members. The survey might be sent to thousands of caregivers identified by geographic information, demographic information, health conditions being managed, age, family situation, and a myriad of other information types.

Interviews have long been another mainstay among techniques for assessing needs (Meister, 1985). Interviews allow the researcher to adapt to the individual situation and probe into specific areas of needs in opportunistic ways not afforded by surveys alone (Kvale, 1996; Weiss, 1993). Although they are very valuable tools, interviews are fraught with difficulties and must be carefully constructed and carefully managed (Meister, 1985; Wilson & Corlett, 1995). The key to the success or failure of interviews is the interaction between the interviewer and the participant. The adaptability and in-depth possibilities associated with interviews can be extremely valuable, but the training of the interviewer, the logistical issues of scheduling, and the resource requirements and limitations associated with one-on-one interviews often make this technique very difficult to implement efficiently. The advantage, however, is that the interviewer can explore specific responses of the person being interviewed such as why some particular approach to preparing for an appointment does not work or how the person would prefer to interact with the doctor beyond the choices provided. Flexibility in pursuing information makes interviewing, whether by telephone or in person, a valuable technique. After a survey of home health caregivers is completed, interviews with a smaller group selected on the basis of their success in providing care might be conducted to get in-depth information on what characteristics of the situation facilitated that success. This in-depth information might be far better collected using interviewing techniques one-on-one than through survey methods.

The use of focus groups is a valuable technique that allows one to realize the virtues of interviewing while interacting with more people at one time than interviewing allows (Greenbaum, 1997; Krueger & Casey, 2000). More importantly, focus groups allow the interaction among the target population that cannot be achieved by either surveys or interviews. Such interactions, facilitated well, can be a tremendously rich source of needs information (Morgan, 1988; Wilson & Corlett, 1995). The example of focus group use by Rogers, Meyer, Walker, and Fisk (1998) cited in the Introduction is an excellent example of the technique's many virtues. There are still difficulties associated with focus groups similar to those found with interviewing, such as facilitator biases and poorly constructed session materials. There also are difficulties unique to focus groups such as the logistics associated with scheduling multiple people in sessions and the threat of poor group dynamics, e.g., a dominant personality type overshadowing other participants.

One approach, developed by IBM, was the Decision Support Center (DSC) concept, which required that instead of communicating verbally, participants typed in responses to questions on computers and then viewed the responses on a screen. This approach allowed for anonymity of responses within the group and corrected for some group dynamic problems. It eliminated the biases or group dynamic problems caused by perceptions of rank or personality differences within the group (M. DiAngelo, personal communication, September 12, 2000).

Another long-used approach involves Delphi techniques whereby responders provide responses on paper, which then are compiled and fed back to the group of respondents for such activities as prioritization or revisions. Delphi techniques, or "slip" techniques, as they are sometimes known, also are used for policymaking and can be applied to a broad range of need assessment topics (Linstone & Turoff, 1984). Respondents often are asked to provide not only a response but also an indication of their level of confidence in their response. In the current example, focus groups might be run that include home caregivers and home care professionals in order to facilitate discussions defining the boundaries between nonprofessional and professional capabilities and limitations and determining how these two "home care" providers groups can better support one another. During the sessions, videotape and written methods are used to capture the discussion and later distill the information into "needs."

To these interactive techniques should be added observational techniques that have long been an accepted and practiced approach to collecting behavioral information on human beings (Meister, 1985; Wilson & Corlett, 1995.) Such techniques have many virtues associated with their direct nature and interpretation. For example, in cases of persons who cannot reliably recount what they do or how they do it, direct observation of their actions can provide an accurate behavioral record for design input. When collected via videotaping, records become long-term data sources that allow for extended post hoc analysis.

Observational techniques, however, can be very effortful and time-consuming. They also entail the risk of misinterpretation when behaviors are incorrectly associated with a specific intent. In some cases, the techniques may not be possible. In the current example, for instance, it might be possible to observe how a patient or home caregiver performs a task in the home using observation, but the other aspects of the care needs may not be appropriate or amenable to observation.

Observational techniques will always be mainstream to human factors design activities. However, the ability to characterize either intentions or cognitive activities such as problem solving or decision-making through observational data is difficult, if not impossible.

Underlying Assumptions

Traditional techniques for conducting needs assessments have been used for decades and continue to be used pervasively. Although useful in allowing system designers to collect the information needed, these techniques are based on a number of assumptions that rarely are checked:

1. *That the sample used is representative of the population.* It is often assumed that the sample selected for surveying, interviewing, focus group inclusion, or observation has sufficient breadth of attributes to represent the population they are trying to understand. The elderly population is one of the most heterogeneous in physical, perceptual, and motor capabilities as well as in demographics.

2. *That needs can be articulated adequately.* The underlying philosophy of these methods is that "people know what they need, so one has only to ask them." Often, needs cannot be expressed clearly or identified clearly at all. Human beings are particularly "blinded" to the diminution of their perceptual and cognitive faculties as they grow older. Individuals, for example, who are not aware that their hearing has diminished, might not articulate the need for hearing aids, no matter how obvious that need is to those around them. Often, the old adage is found to be true that "people don't know what's good for them."

3. *That all needs will be articulated.* It is assumed too often that needs are all viewed as positive and desirable. This often is not the case. For example, how many would expect an elderly sample to articulate yearly driver's testing as a "need" of their age group. The fact is that most of them want the freedom afforded by the privilege of driving for as long as possible and would not articulate as a "need" something they do not want. As a result, needs assessment focusing exclusively on the target population has a built-in bias that must be recognized and addressed.

4. *That observed behaviors represent all pertinent behaviors.* Observation of human behavior is an easy technique to implement, but a very difficult one to record for several reasons. First, if those being observed are aware that they are being observed, their behavior often changes. Second, observers can unintentionally introduce interpretive or categorical biases when recording or classifying the behaviors they observe. Third, if sampling methods are used, there is some probability of missing behaviors because of the nature of sampling itself. Finally, a behavior can be manifested in a variety of subtle ways that are missed easily even when seen.

5. *That needs automatically become requirements representing specifications for a product or system.* This has been discussed previously. Needs do not automatically become or even suggest requirements, much less specifications sufficient to define a product, process, or system. There must be an effortful, defined procedure to support this evolution from needs to specifications. The changes made in 1997 to Code of Federal Regulations 21, Part 820, for the Food and Drug Administration exemplify this. In the changes the FDA began requiring that medical device man-

ufacturers put a process in place to certify that devices had met the needs identified, the requirements developed, and the specifications created for the "intended use" of the medical device being developed.

6. *That if testing shows that the design meets the identified need, then it is efficacious.* There is a circularity in reasoning that if needs are defined and a test conducted to determine whether the product or system meets those needs indicates that it does, then all is right with the world. It is quite a self-fulfilling prophesy that a test designed to focus on a set of predetermined needs finds those needs addressed. The assumption is based on the fact that the test is designed to identify predefined needs. The test or evaluation should allow determination of (a) the needs defined are met, and (b) whether any needs not defined are evident and not met. Typically, tests are not designed to accomplish the latter goal.

The point to be made is not that the traditional techniques should be avoided, but rather that they should not be relied on exclusively to determine and validate "needs." This is especially true in such a critical area as health care for the elderly. Clearly, there are trade-offs that must be made in selecting methods for needs assessment. The choices will depend on many factors including the resources available, the schedule, the nature of the questions, and the availability and accessibility of the population.

Beyond needs assessment, requirements analysis does not appear to have any established techniques associated with it. The House of Quality mentioned earlier is just one example from a wide variety of approaches derived through evolution, iteration, and distillation that follow as many different processes as there are organizations implementing them. With so many assumptions underlying the current set of techniques applied to needs assessment and no standardized and widely accepted approaches for conducting requirements analysis, we need to improve our approach to these critical initial activities (Martin & Bahill, 1996). More effort needs to be expended in establishing standardized techniques for developing requirements and creating specifications from them. What guidance is available can be found in such documents as the IEEE Guide for Developing System Requirements Specifications (1998).

IMPROVING THE APPROACH TO NEEDS ASSESSMENT AND REQUIREMENTS ANALYSIS

It seems clear there must be a renewed appreciation for the criticality of these foundational activities and for innovation in the techniques and approaches we implement to accomplish them. Although the techniques implemented for needs assessment are time-tested and useful, additional techniques are needed to overcome the problems that assumptions may create. The need for innovation in this area is particularly important if a system as complex and crucial as the health care system as it relates to the elderly is to be addressed effectively. Innovation must extend beyond the tools and techniques used. It must involve the very model used to guide current efforts.

Figure 2.1 presents a standardized model for a process involving both needs assessment and requirements analysis. The model recognizes the difference between these activities, discussed earlier, and integrates additional activities to provide a more complete conceptual model of initial system design activities.

It starts with a *comprehensive needs assessment* that brings several sources of information together. These various sources of information help to circumvent and overcome some of the assumptions identified earlier with current approaches to needs assessment and requirements analysis. Four sources of information are identified that makeup the inputs stage of the model.

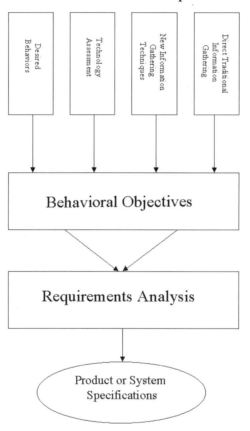

Fig. 2.1   A conceptual model of early system design activities related to needs assessment and requirements analysis.

1. *Direct traditional information gathering.* Despite the assumptions made about traditional techniques (i.e., surveys, interviews, and focus groups), they are ubiquitous and accepted for gathering information. The difficulties associated with them, notwithstanding the need to address them properly, do not outweigh their value for collecting the desired information, and they should be a part of the overall needs assessment. As mentioned earlier, there already are innovations such as the Decision Support Center approach that overcome some of the problems associated with these traditional techniques.

2. *New information gathering techniques.* Many of these techniques are not so much new as they are nonmainstream to needs assessment efforts. They include such techniques as immersive experiential techniques, ethnographic studies, video data analysis, and other emerging techniques and methods.

3. *Technology assessment.* In any effort to conduct needs assessment, there needs to be an understanding of the restrictions or facilitation provided by current and emerging technology. Needs assessment in health for the elderly can be greatly facilitated by such technological breakthroughs as broadband access to the Internet, wireless communication technologies, and telemedical applications. Understanding the factors that influence acceptance of these technologies by the elderly is critical to understanding and overcoming any restriction to use of the technologies. The work cited in this volume by Fisk and Rogers on ATM use by the elderly in this volume, as well as the discussions by Whitten (chap. 7) and Czaja (chap. 9) on telecommunication technologies are excellent examples of this input.

4. *Desired Behaviors.* Design is intended to influence, modify, even control behavior. The ability to "control" behavior through system design, although viewed negatively by some, is valuable. Design used to discourage unwanted behaviors while facilitating desired behaviors, can be a powerful tool. As such, system designers need to identify, wherever possible, user behaviors that are desirable and find ways to design the system to influence these behaviors.

The second stage of the model refers to the development of *behavioral objectives.* These objectives refer to the desired behaviors by users and operators in the system identified by the designers of the system. Such behaviors may be current (e.g., self-examinations) or future (e.g., voice control of medical devices). The development of behavioral objectives has never been a formally defined stage in any needs assessment process, but should be conducted for several reasons:

- Behavioral objectives help to focus the design team on the human element in the system. This element often is the Achilles' heel of large-scale projects because the engineering, programming, or organizational complexities of the system demand such focus that the human being becomes the supposedly adaptable element.

- Behavioral objectives help to identify system needs that often cannot be identified by the users as needs. This is particularly true if the needs are odious to the users, such as examinations or tests considered unpleasant. As mentioned earlier, elderly drivers would not be expected to identify frequent licensing examinations as a need for fear that they might lose their privilege to drive.

- Behavioral objectives can provide continuity between current technological capabilities based on technology assessment and future technological needs for achieving the desired future behavioral objectives. For example, if voice control is desired, it is identified as a future desired behavioral goal for the system.

It is important to note that this stage of the model highlights the added value of including a human factors specialist in the process. With their training across the behavioral sciences and engineering, as well as their human-centric perspective, human factors specialists are uniquely qualified to drive this activity and develop the behavioral objectives of the system. The ability of the human factors specialist to develop accurate and usable behavioral objectives is crucial to any large system design efforts, but has historically been a shortfall of most design efforts. This shortfall may have existed for any number of reasons, but has resulted in large design efforts being driven primary by technology, cost, schedule, or designer experience rather than by an understanding of needs and requirements based on a human-centric focus.

The third stage of the model is *requirements analysis*. This stage is important because it represents the translation of needs assessment information into "what the system must do," which are the data that system designers and developers can understand clearly in their own disciplinary terms. It is, in simplistic terms, deciding how many rooms a house must have and which rooms require plumbing. Requirements analysis, often the stage at which design efforts begin, is not sufficient in and of itself for designing the necessary system because without the needs assessment input, it too often results in a system designed by and for the designers and developers, rather than for the users. As in the given example, the rooms are not based on the user behaviors expected and the needs of those occupying the house.

Requirements, discussed as a definitional issue earlier, are the distillation of needs from many sources into system characteristics and elements that can be implemented and designed by various team players such as engineers, architects, industrial/organizational psychologists, and programmers. The difficulty has always been to find a method for giving all these team members an appropriate and proportional voice. The House of Quality approach mentioned earlier is one documented approach to establishing a formal process, procedure, and structure in which these many voices can be integrated and one set of requirements developed. It is the need for a stable and integrated structure in requirements analysis that often has made large-scale system design difficult, if not impossible. In such a complex system as health care, an additional problem is that, unlike a Disneyland ride, the system cannot be shut down for refurbishing. The need to migrate from the existing system to a new system will be long and arduous and must be planned carefully on the basis of a thorough knowledge of needs and requirements.

The last stage of the model is the *specification stage*. This stage evolves directly from the requirements analysis, but represents the implementation plan. Like a good research write-up, the specification allows any other capable system designer or developer to follow the specification and design a similar system equally successful in meeting the needs of the users. These specifications provide detailed descriptions of implementation, interaction, performance, appearance, information flow, and other characteristics of the initial system design and operations. Several of these specifications at timed intervals represent the migration plan for the system from initial concept to future concept.

The importance of this model lies in its stages. Too often, needs assessment and requirements analysis have been relegated to an afterthought. Then, in hurried fashion, without the proper resources, time, or effort, a poor plan for system design, redesign, or migration has been thrown together. This model represents a four-stage plan for addressing needs and requirements on a very large scale. It is a prerequisite for such a large undertaking as redesigning the health care system to accommodate a changing population with a larger proportion of elderly citizens.

INNOVATIONS IN NEEDS ASSESSMENT TECHNIQUES

There are many innovations for improving the approach to needs assessments and requirements analysis in order to address this particularly crucial and complex system design. The reach of needs assessment needs to be increased dramatically along several dimensions if a redefinition of the health care system to meet the needs of the elderly is to succeed. The expansion must address the breadth, depth, and duration of assessment if the improvements to the approach are to meet the task at hand.

Breadth of assessment can be expanded by studying and including those age groups not yet considered elderly but nevertheless approaching that stage of their life. Cohorts in the 45 to 55 year old and 55 to 65 year old groups need to be included in studies. These are the groups that will become the elderly when the migration of the system is well along in its course. As computer-age cohorts moving into old age, they have a better grasp and facility with current technology and a greater appreciation, on the whole, of what technology can do.

In addition, studies need to include more associated groups. Caregivers, both professional and nonprofessional, must be included if the system is going to be able to expand into its role in meeting the health needs of the elderly over the next 20 years. The needs of the elderly translate into requirements placed on these groups, and their own needs cannot be ignored if the system is to work properly for all. For example, the successful introduction of telemedicine into the home will rely as much on the acceptance of these nonprofessional caregivers as it does on the acceptance of the elderly users.

Depth of assessment can be expanded by studying the ecological factors that influence the system design. Such issues as how the elderly integrate health care into their lives must be better understood, as well as how they deal with the increased intrusion of health care. Adaptive strategies used by the elderly in health-related matters must be identified and incorporated, where possible, into the behavioral objectives and the design. Such strategies represent an excellent example of changes and needs that are not readily identifiable to the elderly themselves. Changes occur over time, often too slowly for clear identification. It is these changes that most often do not get identified by traditional needs assessment methods such as surveys and focus groups. The introduction of new, innovative techniques will be necessary to identify and define these changes and, as a result, design the system to take these changes into account.

Expanding the horizon view or duration of assessments is critical and suggests the need for more longitudinal studies. Studies such as those conducted on heart disease in Framingham, MA (Kannel, 1990) and the Baltimore Longitudinal Study on Aging begun in 1958 (Andres, 1979; Shock et al., 1984; Verbrugge, Gruber-Baldini, & Fozard, 1996) are outstanding examples of the type of effort needed. While these studies focused on areas of medical etiology and the biological impact of aging, they provide illustrious examples of the value that such long-term study provides for understanding complex interactions and effects. Such studies can positively impact both design and evaluation by providing ongoing data about how people use and adapt to the health care system as they age and what metrics can provide the best feedback on the quality, efficiency, and effectiveness of the health care system for the elderly (Menard, 1991).

Emerging and Innovative Techniques in Needs Assessment and Requirements Analysis

Although traditional techniques for needs assessment and requirements analysis will likely continue to be a mainstay in the up-front data collection effort, there are innovations and capabilities available to expand on the data sources used and to improve the quality of the data collected. Some of these techniques are not new, but have been used only rarely by specific disciplines. Other techniques are a direct outgrowth of new technologies that are emerging and evolving faster than they can be implemented in research. The following approaches represent some innovative techniques that can be applied to the problem by the research, design, and development community.

Immersive-Experiential Techniques. Not since Watermelon Man, the 1970s dark comedy about a White bigot turned Black and his experiences of bigotry and racial hatred, has immersive-experiential–based research been showcased to the extent demonstrated by Dr. Patricia Moore from 1979 to 1982. Dr. Moore, now a principal at Bresslergroup, Inc. in Phoenix, AZ, as a graduate student in industrial design, spent 4 years traveling through 116 cities in the United States and Canada dressed and outfitted as an elderly woman. Her intent was to experience the life of the elderly. To accomplish this, she not only dressed and made herself up to appear as an 80-year old woman, but also bound her joints, wore glasses that diminished her sight, and reduced her hearing ability with ear plugs to emulate the aged character she portrayed. Her "in situ" research and the knowledge she gained regarding the interactions of the elderly with both society and design provide an important example of using innovative techniques to understand what it means to be elderly. Her findings and experiences are presented in her 1985 book entitled *Disguised: A True Story* (unfortunately out of print).

This approach is now used sometimes by product development companies in product evaluations when test participants wear reading glasses smudged with Vaseline to reduce their vision in reading labels or wear gloves while performing a task such as opening a bottle or using a common household tool. The author himself was able to appreciate Dr. Moore's immersive-experiential approach once when he was blinded by conjunctivitis and rendered blind for several days, requiring that his other senses and friends do his work.

The use of immersive-experiential techniques to "walk a mile in the shoes of someone else" is a tremendously effective approach to developing a sense of one's users. There is no observational or conversational equivalent. Although the author has heard anecdotally about the use of this technique by some geriatric and gerontological training programs, it could have tremendous value if used more by system designers and developers to "understand what their users experience." For some researchers, this qualitative approach will not produce the same quantitative dataset they seek for analysis and interpretation. The nature of this technique, however, through personal experience can result in a personal epiphany and profound understanding of the users for whom the system is being designed.

Internet-Based Tools. The emergence of the Internet and the World Wide Web have led to an explosion of possibilities in data collection. The Internet provides the ability to collect information on behaviors, preferences, needs, activities, opinions, and change in ongoing and cost-effective ways as never before. Several opportunities have been explored over the past 10 years:

- Collecting data directly relevant to needs of the elderly. One example of this is The Mature Mart (www.maturemart.com), an interactive website selling products specifically for seniors. Data are collected from this Web site to determine future product needs. The use of such kiosks as collection points for "needs" information could provide a long-term ongoing source of information for system designers in the health care area.

- Internet data collection tools have been developed and are in use on a regular basis now. These tools are used to collect massive amounts of data through Internet surveys. The technology has reached the point of allowing researchers to form virtual focus groups through emerging meeting and conferencing tools over the Internet. Geography and the postal system are no longer boundaries to the rapid collection of data. Although acceptance of the technology is lower for the elderly, the ability of researchers and system designers to collect data from a broad set of representative users has never been greater. The potential future return of Internet-based surveys from cohorts now in 45-to-55-year-old and 55-to-65-year-old age brackets could have very large returns in redesigning of the health care system to serve the elderly better.

- Kiosk-based data collection networks and their capabilities also are emerging in stores and locations such as Disney World/Epcot. Mature Mart has more than 800 kiosks in stores through which the elderly can purchase products, respond to surveys, or make requests, suggestions, or comments. These systems could be pervasive and continuous sources of data on the elderly relative to health care issues or a broad range of other issues.

The Internet and World Wide Web can be viewed as tools in redesigning and redefining the health care system for the elderly. The Internet can be conceptualized as a communication channel for exchanging information and collecting data at an unprecedented rate, unbounded by geography or traditional process modes, and in innovative ways such as electronic surveys. The World Wide Web must be approached more carefully as suggested by Stronge, Walker, and Rogers in Chapter 14 of this volume. It is a powerful tool that can quickly overwhelm anyone looking for information.

Wireless Telecommunications. Where computer-based Internet capabilities leave off, wireless telecommunications pick up. The ability now to interact with the various user groups to understand their needs and behaviors is almost unbounded. A group of representative users can now be equipped to time sample their activities, inquire about their choices, collect their opinions and recommendations, and solicit their feedback and impressions on an almost continuous basis. The use of cell phones and personal devices such as personal digital assistants (PDAs) allows access to individuals as never before. The cost of such devices is dropping as the capabilities, bandwidth, and coverage are expanding.

Wireless communications and teleconferencing capabilities provide us with the opportunity to conduct a widespread longitudinal study of needs, behaviors, attitudes, and activities using a virtual communications model as the structure, and to establish a very efficient data collection method in the process.

New Research Methods for Identifying Needs Indirectly. New methodologies have emerged in the past 5 years such as cognitive task analysis (CTA); (Gordon & Gill, 1997), naturalistic decision-making (NDM); (Klein, Orasanu, Calderwood, & Zsambo, 1993), and situational awareness (SA); (Endsley, 1995), which allow research and examination of human behavior from a cognitive perspective in all its complexity. These techniques, if applied in an ongoing manner to a broad set of cohorts between 45 and 85 years of age, can allow an understanding of the changes occurring in the elderly that have an impact on the effectiveness of the health care system in meeting their needs.

Cognitive task analysis allows an understanding of the cognitive processes involved in specific task activities and to understand how they change with age. Naturalistic decision making provides insight into decisions made in real-world contexts under real-world dynamic conditions. Situational awareness reflects the contextual knowledge of individuals such as the elderly relative to dynamic situations in which they must function and the extent to which they are aware of the various elements and information around them at any given point in time.

In contrast to many years of observing behaviors and asking questions that provided limited understanding of the cognitive processes underlying them, these methods have expanded the ability to interpret what underlies human behavior. Furthermore, and just as importantly, techniques such as naturalistic decision making allow researchers to work in situ and not be constrained to the laboratory. To meet their purposes in assessing needs and requirements for the health care system for the elderly, this ability is vital. Clearly, in health care, the interactions among the many players and the informational complexity and dynamics need to be evaluated as they relate to needs and requirements of the elderly. Although some of these interactions can be evaluated in the laboratory, they are most realistically understood in the context of the real world.

Ethnographic Studies and Video-Data Analysis. The need to evaluate needs and requirements in an ecological sense requires the ability to collect data and interpret it in situ. Ethnography, the study of behaviors with a cultural context, has been expanded over the past few years to include environment as well as culture (Denzin, 1996; Hammersley & Atkinson, 1995; Kutsche, 1997). The techniques are used at both the individual level and cohort levels with data, both visual and nonvisual, being collected in such diverse settings as homes, nursing homes, hospitals, shopping malls. The use of visual (e.g., video, photographs, drawings) as well as nonvisual sources of data is important because it adds both qualitative (content) and quantitative (numerical) data to the set.

Video data analysis is increasingly important as the tools for analyzing the video grow stronger. The ability to work in digital modes has been a breakthrough for such analysis. Tools are being developed that allow not only digital processing, but also multiple video streams and post hoc synchronization of video and nonvideo data streams (e.g., physiological data, verbal data, and operational data). It has long been known that video is the richest source of data with many virtues such as unlimited post hoc analysis opportunities, high archival value, and tremendous impact on system design and development teams to influence design decisions. The problem has always been the tedious nature of analysis, the time required to extract data, and the legal issues associated with permission to videotape human beings (a problem still to be addressed). The potential of video-based data as a method of collecting needs-related information is huge, but must overcome obstacles relative to specifying content and extracting and analyzing it before that potential can be realized.

THE ROLE OF HUMAN FACTORS

Human factors focuses on the design of human interfaces between technology on the one hand and people on the other. The importance of this role has long been recognized and emphasized for health care (Klatzky & Ayoub, 1995). Through their understanding of human capabilities, limitations, behaviors, and life changes among these characteristics, human factors specialists design systems, including medical systems, that adapt to people instead of requiring people to adapt to systems. Human factors specialists are uniquely trained to collect, analyze, and interpret data associated with needs and requirements because their perspective, as represented throughout this volume, is "human-centric" in nature, amalgamating human behavior with technological engineering. Furthermore, human factors specialists are trained to integrate both qualitative and quantitative data from multiple sources, interpret the results, and generate both system and detailed level recommendations for meeting the needs of the human beings in the system.

Human factors specialists also are trained to work on multidisciplinary teams involving a range of skills and knowledge such as engineering, behavioral science, communication, medicine, and programming. Human factors specialists are trained to facilitate interactions among the members of such teams. In a system as complex as the health care system, the number of members on the various teams designing a new system will be very large and the variety of disciplines represents extremely broad. The ability of these disciplines to communicate effectively and reach the inevitable compromises required will be facilitated by the involvement of human factors specialists whose orientation and perspective are the result of interdisciplinary training and experience.

CONCLUSION

Addressing the health care needs of the elderly in the new millenium is vital to the health of all whether as individuals, the nation, or the world. The foundation on which that design is based and the ability of that design to work effectively depend on the quality of the data and the extent of the effort to assess the needs and determine the requirements of such a system. This is a massive undertaking. If the delays and failures associated with upgrading the air traffic control system in the United States are any indication, this undertaking is larger than current approaches can handle.

This discussion strives in a very small space to suggest that the tools and techniques capable of achieving the level of needs assessment are emerging if they only are used. The standardized tools and processes for requirements analysis still need to be developed and refined further to ensure translation of needs into specifications for redesigning the health care system. In light of this, the research, design, and development community is challenged to focus on a model and tools that will allow massive amounts of convergent data to be collected, compiled, and interpreted. These data must be compiled into the needs, behavioral objectives, requirements, and, ultimately, system specifications that will migrate the current health care system to a new, improved system capable of proving effective and efficient for the health care needs of a more elderly population in the new millennium.

REFERENCES

Andres, R. (1979). *The normality of aging: The Baltimore longitudinal study*. (NIA Publication No. 79–1410) NIA Science Writer Seminar Series. Washington, DC: U.S. Department of Health, Education, and Welfare.

Code of Federal Regulation, 21 (1997) Parts 800 to 1299, Washington DC: Office of the Federal Register.

Denzin, N. K. (1996). *Interpretive ethnography: Ethnographic practices for the 21st century*. Newbury Park, CA: Sage.

Dillman, D. A. (1999). *Mail and internet surveys: The tailored design method*. New York: Wiley.

Endsley, M. A. (1995). Toward a theory of situation awareness in dynamic systems. *Human Factors, 37*, 85–104.

Gordon, S. E., & Gill, R. T. (1997). Cognitive task analysis. In C. Zsambok & G. Klein (Eds.), *Naturalistic decision making* (131–140). Hillsdale, NJ: Lawrence Erlbaum Associates.

Greenbaum, T. L. (1997). *The handbook for focus group research*. Newbury Park, CA: Sage.

Hammersley, M., & Atkinson, P. (1995). *Ethnography: Principles in practice*. New York: Routledge.

Hauser, J. R., & Clausing, D. (1988). The house of quality. *Harvard Business Review, 30*, 7–18.

Howell, W. C. (1997). Foreword, perspectives, and prospectives. In A. D. Fisk & W. A. Rogers (Eds.), *Handbook of human factors and the older adult* (pp. 1–6). New York: Academic Press.

Institute of Electrical and Electronic Engineering, *IEEE guide for developing system requirements specifications* (1998). Piscataway, NJ: IEEE.

Kannel, W. B. (1990). Contribution of the Framingham Study to preventive cardiology. (Bishop Lecture). *Journal of the American College of Cardiology, 15*, 206–211.

Klatzky, R. L. & Ayoub, M. M. (1995), Health care. In R. Nickerson (Ed.) *Emerging needs and opportunities for human factors research*. (131–157). (National Research Council Report). Washington, DC: National Academy Press.

Klein, G. A., Orasanu, J., Calderwood, R., & Zsambok, C. E. (Eds.) (1993). *Decision-making in action: Models and methods*. Norwood, NJ: Ablex.

Krueger, R. A. & Casey, M. A. (2000). *Focus groups: A practical guide for applied research*. Newbury Park, CA: Sage.

Kutsche, P. (1997). *Field ethnography: A manual for doing cultural anthropology*. Newbury Park, CA: Sage.

Kvale, S. (1996). *Interviews: An introduction to qualitative research*. Newbury Park, CA: Sage.

Lee, E. S., Forthofer, R. N., Lorimor, R. J. (1990). *Analyzing complex survey data*. Newbury Park, CA: Sage.

Linstone, H. A., & Turoff, M. (1984). *The delphi method: Techniques and applications*. New York: North-Holland.

Martin, J. A. & Bahill, A. T. (1996). *System engineering guidebook: A process for developing systems and products, Systems Engineering Series*. Boca Raton, FL: CRC Press.

Meister, D. (1985). *Behavioral analysis and measurement methods*. New York: Wiley.

Menard, S. (1991). *Longitudinal research*. Newbury Park, CA: Sage.

Moore, P. (1985). *Disguised: A true story*. Waco, TX: Word Books.

Morgan, D. L. (1988). *Focus groups as qualitative research*. Newbury Park, CA: Sage.

Rea, L. M., Parker, R. A., & Schrader, A. (1997). *Designing and conducting survey research: A comprehensive guide*, Jossey-Bass Public Administrative Series. San Francisco, CA: Jossey-Bass.

Rogers, W. A., Meyer, B., Walker, N., & Fisk, A. D. (1998). Functional limitations to daily living tasks in the aged: A focus group analysis. *Human Factors, 40*, 111–125.

Shock, N. W., Gruelich, R. C., Andres, R., Arenberg, D. Costa, P. T. Jr., Lakatta, E. G., & Tobin, J. D. (1984). *Normal human aging: The Baltimore longitudinal study*. (NIH Publication No. 84–2450). Washington, DC: U.S. Department of Health and Human Services.

Verbrugge, L. M., Gruber-Baldini, A. L., & Fozard, J. L. (1996). Age differences and age changes in activities: Baltimore longitudinal study of aging. *Journal of Gerontology, Social Sciences, 51B*, 830–841.

Warwick, D. P. & Lininger, C. A. (1975). *The sample survey: Theory and practice*. New York: McGraw-Hill.

Weiss, R. S. (1993). *Learning from strangers: The art and method of qualitative interview studies*. New York: Free Press.

Wilson, J. R., & Corlett, E. N. (Eds.) (1995). *Evaluation of human work: A practical ergonomics methodology*. Bristol, PA: Taylor & Francis.

PART II

COGNITIVE ISSUES

3

Consideration of Memory Functioning in Health Care Intervention With Older Adults

Anderson D. Smith
Georgia Institute of Technology

Two of the most apparent changes with aging are increased need for health care and decreased memory functioning. The exact relation among health, memory, and aging, however, is a complex one. Some diseases associated with advanced age are neurological in nature and primarily characterized by memory loss (e.g., Alzheimer's Disease). With diseases other than neurologic pathologies, however, it is not at all clear whether there is a direct relation between health and memory functioning (Deeg, Kardaun, & Fozard, 1996). For example, it is known that certain chronic diseases are associated with greater memory change (e.g., cardiovascular disease; Schaie, 1996). That is, those who have the disease show greater memory loss than those who do not. There is still considerable debate, however, as to whether the disease state causes cognitive declines or whether greater cognitive declines produce poorer adaptation to environmental stressors, adaptation that is necessary to maintain health and wellness. Schaie (1996) pointed out that more cognitively competent individuals may be more able and likely to engage in the appropriate behaviors that postpone the onset of serious disease.

One popular research strategy for looking at the relation between health and memory is to use causal modeling with these variables. Measures of self-reported health are collected at the same time memory is tested in the laboratory. In one such study (by Earles, Connor, Smith, and Park, 1997), for example, several measures of self-reported health and memory performance were collected in a large number of adults 20 to 80 years of age. The health measures were adapted from the Duke University (1978) Older American Resources and Services Multidimensional Functional Assessment Questionnaire. The questions were (a) "How would you rate your present health?" on a scale of 1 (poor), 2 (fair), 3 (good), or 4 (excellent); (b) "How much do health troubles stand in the way of you doing things you want to do?" and (c) "How many prescription medications are you presently taking?" The memory measures included different tests of three commonly researched memory tasks: working memory, free recall, and cued recall. Working memory involves the ability of the adults to engage in "online" information processing. When hearing and comprehending either conversation or text, working memory capability allows an individual to encode and store information as he or she is comprehending it. Free recall and cued recall tasks simulate one's ability to remember past experiences directly via long-term memory. Covariance structure modeling of the data is used to indicate the best-fitting model of the relations among the variables. For example, the model would show whether there was a direct relation between two variables such as health and memory, or whether the relation was indirect, mediated through some third variable.

The Earles et al. (1997) study supported a model showing health mediating the relation between age and memory, but only through another cognitive measure of perceptual speed. Perceptual speed indicates how fast an individual can make perceptual judgments such as whether two letter strings (e.g., ckhjlth) are the same or different. Figure 3.1 shows this model. Age was directly associated with decreased health and slower perceptual speed, but the relation between aging and the memory measures was indirect, mediated only through perceptual speed.

Path coefficients are from LISREL's completely standardized solution, and all paths with a ± value less than .05 are significant. The path coefficients (numbers in the model) indicate the strength of the relation being modeled. In this study, as in others using similar modeling techniques (Earles & Salthouse, 1995; Hultsch, Hammer, & Small, 1993; Luszcz, Bryan, & Kent, 1997; Stankov & Anstey, 1997), mediated effects of health on memory are found, but not direct effects of health on age-related memory performance. Instead, health has its effects on more elementary cognitive processes such as perceptual speed or sensory perceptual functioning that in turn influence memory performance. Some have argued that measures of speed or sensory function reflect core measures of the efficiency or integrity of information processing in the brain (Baltes & Lindenberger, 1997), and it appears that health has its effect on these types of measures rather than on memory or complex cognition directly.

On a more cognitive level, many have argued that these elementary measures (e.g., speed or sensory functioning) are estimates of the cognitive resources available to perform cognitive tasks (Zacks, Hasher, & Li, 1999). The fact that older adults are slower and have diminished working memory capacities means that they have fewer cognitive resources to allocate to self-controlled processing in cognitive tasks such as deliberate remembering.

Whatever the interpretation, most of the literature shows that these mediated effects of self-reported health on elementary processes are very small and account for only a very small part of age-related variance in memory performance. On the other hand, as the Earles et al. (1997) study showed, the elementary processes themselves account for a large proportion of the age-related differences in memory

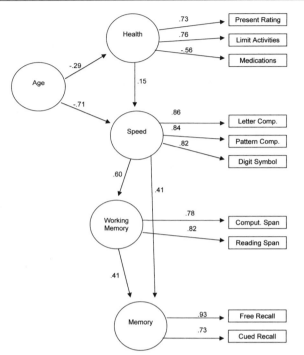

FIG. 3.1. Best-fitting model of health, age, and memory. Adapted from "Interrelations of Age, Self-Reported Health, Speed and Memory," by J. L. K. Earles, L. T. Connor, A. D. Smith, & D. C. Park, 1997, *Psychology and Aging, 12*, p 681.

performance. The proportion of age-related variance in memory accounted for by the indirect effects of health, however, is very small.

Regardless of the relation among age, health, and memory, the fact remains that two major effects of aging are memory change and a greater susceptibility to health problems. The correlation is present whatever the causal relation, so it is critical that age-related memory functioning is understood in the design and consideration of health care interventions that will be used with older adults.

Many recent reviews of the aging and memory literature exist (Craik & Jennings, 1992; Light, 1996; Smith, 1996; Zacks, Hasher, & Li, 1999). In general, these reviews provide consensus as to the major conclusions from this literature. Table 3.1 lists seven general conclusions that can be drawn from examination of the basic, theory-based research on aging and memory. Although this list certainly does not come close to exhausting research findings, it does provide some general principles about aging and memory that are useful in thinking about health interventions with older adults. In recent reviews of applied cognitive aging research, Park (1992) and Rogers and Fisk (2000) have suggested that theory-based research on cognition and aging has now developed sufficiently so that solutions to real-world problems are beginning to be framed by that research.

It is apparent from examining this list of conclusions that the best description of memory functioning in people who are aging is not one of simple decline. Although many aspects of memory do decline, other memory functions are maintained and even improved into late adulthood. It is this complexity of loss, maintenance, and gain that provides a framework for application and intervention. Interventions that require older patients to "remember" information should avoid memory situations for which research and theory indicate significant loss by older adults, and instead emphasize, if possible, memory situations for which invariance or even gain with age is found.

SELECTIVE OPTIMIZATION WITH COMPENSATION

The preceeding advice is the foundation for the model of "successful aging" proposed by Baltes and Baltes (1990) called "selective optimization with compensation" (Fig. 3.2). According to this model, successful aging is controlled by three adaptive processes operating in the context of the conditions associated with aging: loss of specific functions, continued adaptation to change, and reduced reserve capacity.

The first adaptive process involves an age-related increase in specialization (selection) of cognitive resources and skills. In other words, with aging there is a decrease in the number of memory processes and skills that remain efficacious. These

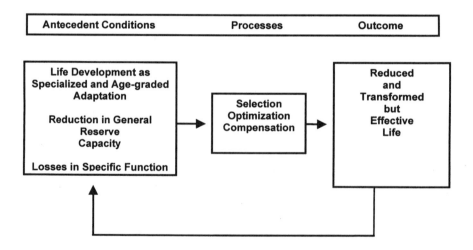

FIG. 3.2.   Selective optimization with comprehension. Adapted from "Psychological Perspectives on Successful Aging by P.B. Baltes & M.M. Baltes. In P.B. Baltes & M.M. Baltes (Eds.), Successful aging: Perspectives from the Behavioral sciences. New York: Cambridge University Press, 1990, p.11. Copyright ©1990 by the authors and Cambridge University Press. Reprinted with permission.

processes and skills therefore must be "selected" or emphasized to minimize problems of remembering in older adults.

Second, as individuals age they can continue to engage in behaviors that improve reserve capacity (optimization), especially with memory and knowledge skills. Evidence suggests that most memory skills can be improved by further learning, and optimization involves enriching and augmenting memory skills through practice and training.

Third, compensation can take place through change in strategy, use of mnemonics, or reliance on different ways of maintaining overall memory performance. *Selection* involves emphasizing memory skills that show age invariance or improvement and de-emphasizing skills that show decline. *Optimization* involves continued improvement of memory skills through training and practice. *Compensation* involves completing the memory requirements differently through strategy change or by using different memory processes. Selection and compensation are possible because memory changes with aging are complex, involving components of memory function that show different patterns of loss, gain, and maintenance. Optimization is possible because there is a great deal of plasticity in most behaviors, and individuals are not always performing at the limit of their reserve capacity.

In summary, intervention strategies can emphasize memory functions that do not change (selection), improve memory functioning through training or extra practice (optimization), or find new ways to satisfy the memory requirements of the task (compensation). Empirical evidence shows that the self-reported use of selection, optimization, and compensation by older adults is correlated with subjective indicators of successful aging such as well-being and positive emotions (Freund & Baltes, 1998).

CONCLUSIONS ABOUT AGING AND MEMORY

The seven conclusions in Table 3.1 are broad, encompassing a great deal of research on aging and memory. They represent principles about aging memory that can be used in developing interventions to be used with older adults for tasks that require memory.

A distinction often is made between episodic (context-cued) memory and semantic (concept-cued) memory (item 1 in Table 3.1). Episodic remembering requires reinstating the context of the original experience in order to remember. Examples involve the need to consider the following questions: "Did I take my medicine this morning?" "What was that doctor's name who examined me yesterday?" On the other hand, semantic memories consist of general knowledge or knowledge associated with some specific domain of individual expertise. Examples involve these considerations: "Heavy salt use is not good for my blood pressure." "The pharmacist would have a new thermometer to replace my broken one." Older adults show increasing problems with episodic recall but only minimal problems with semantic recall. In fact, some studies show increases in semantic memory throughout the life span well into late adulthood. In a large-scale study examining predictors of memory performance, for example, Park et al. (1996) found significant decreases in episodic memory (free recall) across the life span (Y mean, 26.2; O mean, 21.1) and significant increases in semantic memory (vocabulary) in the same study participants (Y mean, 22.6; O mean, 27.7).

Because episodic recall is context dependent, retrieving episodic information is easier if the context is well integrated with the to-be-remembered material (item 2 in Table 3.1). In a study by Smith, Park, and Whiting (1998), related cue–target pairs of words (ring, finger) were presented with two additional context words that either facilitated the relation between the cue and target (gold, wedding), or were neutral (tree, flower). When the context words were related to the cue–target pair, memory was improved for the target word when later presented with the cue.

In fact, mnemonic training is essentially based on the preceding logic. Most mnemonic techniques typically provide a well-learned and integrated set of contextual cues to support remembering a set of episodic information, and this training does work with older adults (Yesavage, Lapp, & Sheikh, 1989).

Whereas a facilitating context can support episodic memory in all age groups, a distracting or irrelevant context is especially detrimental to older adults (item 3 in Table 3.1; Earles, Smith, & Park, 1994; Kausler & Kleim, 1978; Park, Smith, Dudley, & Lafronza, 1989). In fact, one specific theoretical view of aging and memory is related to this notion. Hasher and Zacks (1988) attributed age-related differences in memory to a decline in attentional inhibitory control over the contents of memory. According to this view, inhibitory failures of older adults increase the presence of irrelevant memory content that increases interference at both encoding and retrieval.

TABLE 3.1.

Seven Conclusions About Aging and Memory

1. Aging involves loss of "episodic" memory, but maintenance of "semantic" memory. (Older adults may not remember whether they took their medicine that morning, but would remember that "medicine X" treated their health problem.)

2. Age differences in episodic memory are reduced if contextual cues are integrated with to-be-remembered information. (Remembering to take one's medicine in the morning is improved if the pill bottle is placed next to the toothbrush.)

3. Older adults are more affected by distracting context when trying to remember episodic memories. (Having the daughter visit with the grandchildren is likely to interfere with the older adult's ability to maintain the medication regimen prescribed for the day.)

4. While there are age-related differences in explicit memory, there are minimal changes in implicit memory. (The older adult will not remember the name of the doctor she visited yesterday, but later produces the name unconsciously when discussing a neighbor with the same first name.)

5. There are age changes in deliberate recollection, but not with memories based on habits or familiarity. (Once well-learned habits are developed, remembering those medical appointments becomes easier for the older adult.)

6. Aging reduces processing capacity as measured by working memory. ("Understanding complicated medical instructions is not as easy as it was when I was younger.")

7. Age differences in memory increase with requirements of more self-initiated processing, but decrease with more environmental support. (Placing that medication schedule on the refrigerator really helped me remember to take my medicines throughout the day.")

In the study mentioned earlier, Smith, et al. (1998), also presented context words that were incongruent or distracting to the relation of the cue–target pair (ring, finger). The distracting context words (phone, bell) were related to another meaning of the homograph cue word, and thus were incongruent to the relation between the cue and the target. While both sets of extra words, facilitating (gold, wedding), as well as distracting (phone, bell) were related to the homograph cue "ring," the distracting words interfered with the integration of the cue and target. As compared with neutral words, the facilitating words benefited both young and old alike (young, 22.6% facilitation of performance; old, 23.3%). The distracting words, however, significantly reduced the memory performance of only the older group (young, 1.6% reduced performance; old, 38.7%). As Park (1992) pointed out in her review of applied cognitive aging research, the implications of greater distraction and interference in older adults in the design of labels, instructions, and text material is substantial. Older adults need to be focused better on relevant aspects of the task.

Another distinction in memory is between explicit memory and implicit memory. Whereas conscious awareness is a part of the remembering process for explicit memories, implicit memory effects can also be demonstrated for which there is no conscious awareness. Although there are large reliable differences between age groups in explicit episodic memory, there are essentially no such differences in implicit memory (item 4 in Table 3.1) (see Howard, 1996, for a review). One of the earliest studies looking at implicit memory demonstrated this finding in different adult age groups. Light and Singh (1987) first gave young and old participants 20 words and asked them to rate the "pleasantness" of the words. Then later, half the participants were given the first few letters the words (word stems) and asked to complete the words with those presented in the earlier list (i.e., explicit recall). The other half of the participants were not told that the test was a memory test. They were instructed simply to complete the word stems with the first word that came to mind (i.e., implicit recall). As has been found frequently in the literature on aging memory, younger adults were much better than older adults when recalling with the explicit instructions (young, 55% recalled; old, 28%), but did not differ from the older adults when the implicit instructions were used (young, 28% completed; old, 20%).

Some have argued that the age differences with explicit memory and age invariance with implicit memory are produced because two distinct memory systems are involved (Tulving & Schacter, 1990). Another view is that in any memory task, remembering involves a combination of deliberate recollection (intentionally trying to remember) with established and familiar habits (just knowing). Remembering can be intentional, effortful, and deliberate; automatic and based on habit; or both. Jacoby (1991) and his colleagues, for example, have extensively used a new "process dissociation" procedure to estimate remembering by habit and recollection. In all of these experiments, age differences are found with the recollective component of remembering, but not with remembering based on habit (item 5 in Table 3.1).

For example, Hay and Jacoby (1999), using the process dissociation procedure, were able to show that older adults were impaired in their ability to engage in recollection, but did not differ from younger adults in their use of habit (memory based on familiarity). Using homographs, like those used in the distracting context study mentioned earlier (Smith et al., 1998), words with two meanings (e.g., organ–music; organ–heart) were selected. In training, the words were presented 75% of the time with one meaning (typical) and only 25% of the time with the other meaning (atypical). The training was done to establish a habit for one of the word meanings. Then in a study trial the researchers presented the participants with either the typical (based on habit) or atypical (inconsistent with the habit) word as the response. In a later test, the participants were asked to recall the word that was paired with the cue during the study trial. During congruent study trials, the participants studied words that were made typical (P[recall] = P[recollect] + P[habit (1-R]). During incongruent trials, they studied words that were atypical. In the congruent trials, the participants could give the correct word either by recollection or habit. However, in the incongruent trials, the participants studied words that were atypical in terms of their established habit. Habit or familiarity was shown to be a source of error, so the probability of incorrectly producing a typical response must be based on habit (P[recall] = P[habit(1-R]). Subtracting the probability of an error during incongruent trials from the probability of a correct response during congruent trials provides an estimate of recollection (P[recollection] = congruent - incongruent).

As can be seen in Table 3.2, the probability of responding with a typical item in the incongruent condition is greater for older adults. They made more memory slips basing their performance on habit. The superior performance of the younger adults therefore was based on their higher recollection scores. There were significant age differences in estimates of recollection, but not of habit. The implication of this finding is clear: The more that the memory task can be controlled by habit rather than deliberate recollection, the better the older adults will do in performing the task.

TABLE 3.2.

Probability of Responding With a Typical Item, and Estimates of Recollection and Automatic Influences as a Function of Age and Distinctiveness

Age and Condition	Trial Type		Estimates	
	Congruent	Incongruent	Recollection	Habit
Young				
Nondistinctive	.84	.41	.44	.72
Distinctive	.88	.29	.60	.70
Old				
Nondistinctive	.80	.51	.29	.72
Distinctive	.80	.50	.30	.72

Note. Adapted from "Separating habit and recollection in young and older adults: Effects of elaborative processing and distinctiveness," by J. F. Hay and L. L. Jacoby, (1999), *Psychology and Aging, 14*, p. 131. Used with permission of authors and the American Psychological Association.

Self-report procedures support the same conclusion. Parkin and Walter (1992), for example, had different aged adults indicate whether recognition of words from an earlier study list was based on "remembering" (recollection) or on "knowing" (familiarity). A "know" judgment indicated recognition of the word but no remembered information about its occurrence on the list. If the original study list occurrence was recollected, however, the appropriate response would be "remember." Whereas young adults indicated that more than 50% of the recognized words were remembered, old adults indicated "remember" for less than 20% of their recognized words. One the other hand, young adults responded "know" to only 25% of the recognized words, whereas old adults indicated "know" to more than 40% of the words.

One way to reduce the degree of deliberate recognition is to increase the familiarity of the task. For example, extensive practice with the task often can reduce the requirements for controlled processing, producing automaticity of responding (Park, 1992). Whereas is not yet clear whether older adults can reach the same level of automaticity as do younger adults (Rogers, 1992), it is clear that the controlled processing requirements of a task can be reduced with practice (Clancy & Hoyer, 1994)

The fact that older adults have trouble when intentionally and deliberately processing information is most likely because of the reduced resources available to older adults (Craik & Byrd, 1982). Intentional or controlled processing during a memory task requires that cognitive resources be devoted to the task. When elementary measures of processing resources (e.g., speed or working memory) are measured at the same time memory is tested, the processing resource measures account for a vast majority, if not all, of the age-related variance on the memory task (item 6 in Table 3.1). Using structural equation modeling, for example, Park et al. (1996) found that when speed and working memory were included in the model, all of the age-related effects on memory went through these two variables. Figure 3.3 shows that the best-fitting model involves no direct effect of age on spatial recall, free recall, or cued recall when speed and working memory are considered as moderating variables.

The clear implication of the fact that most age-related differences in memory can be explained for by processing resources is that interventions should be designed that require minimum processing requirements to perform the memory task. This can be done by using training or extensive practice that optimizes memory performance by increasing automaticity and the extent to which the task is governed by habit rather than deliberate processing.

Another way of reducing the requirements for self-initiated, deliberate processing is to design the task itself to support memory and reduce reliance on controlled processing. Craik and Anderson (1999) suggested that there is a direct relation between the amount of deliberate processing required in a memory task and the environmental support provided by the task itself. Age differences are small if environmental support for the task is high, and age differences are large if environmental support is small (item 7, Table 3.1). According to Craik and Anderson (1999), environmental support includes (a) the guidance of processing when the information is to be encoded and retrieved, (b) the easy availability of relevant prior knowledge that could support encoding and retrieval (integration with past knowledge), and (c) the presence of actual cues that could aid encoding and retrieval (Zacks, Hasher, & Li, 2000). For example, recognition memory tasks produce

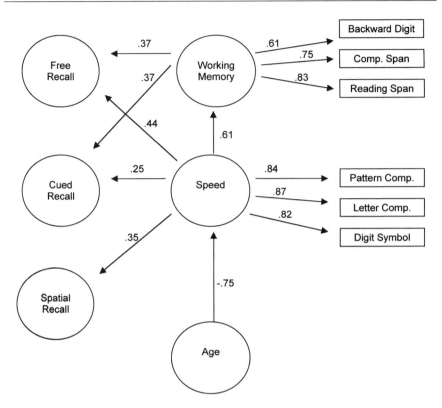

FIG. 3.3.   Best fitting model of age, speed, working memory, and episodic recall. Adapted from "Media-tors of Long-Term Memory Performance Across the Life Span," by D. C. Park et al., 1996, *Psychology and Aging, 11,* p. 631. Used by permission of authors and the American Psychological Association.

smaller age differences than recall tasks because recognition tasks provide the opti-mal cues for environmental retrieval support (the words themselves). Using a sec-ondary task procedure, Craik and McDowd (1987) showed that recall tasks require more processing resources than recognition tasks, as estimated by the disruption by the second task, or the cost of having to perform it. Recognition provides more envi-ronmental support, requires less deliberate processing, and produces smaller age differences.

To summarize Table 3.1, age differences are greatest when episodic recall is re-quired and when cues for episodic recall are distracting, or when they are facilitative but not well integrated with the to-be-remembered information. Age differences are small, however, when the task does not involve deliberate recollection, but is based on familiarity or implicit memory. Processing requirements for memory can be reduced, and thus age differences reduced, by making the responses more auto-matic (i.e., well learned skills or familiar knowledge), or by reducing the processing requirements of the task through environmental support.

EXAMPLES OF THEORY-BASE INTERVENTIONS

This volume is filled with examples of theory-based and research-based health care interventions. For the sake of illustration, however, the examples that follow deal with the use of the conclusions mentioned earlier and health care issues involving with (a) medication adherence, a topic that has generated a great deal of cognitive aging research, and (b) the design and use of medical devices and new technologies, an area not well researched.

Medication Adherence

As mentioned, considerable theoretical research exists on the memory factors that are important in medication adherence (see Park & Jones, 1997, for a review). Instead of merely saying that failure to adhere to a medication regimen is a "memory" problem, Park and Jones (1997) have actually determined what memory characteristics are responsible for failure to adhere. In one study of hypertensives, for example, Morrell, Park, Kidder, and Martin (1997) showed that working memory capacity was related to medication adherence, whereas prospective memory (remembering to do something in the future) and long-term memory abilities (trying to remember something from the past), intuitive candidates for failure to take medications, did not predict adherence as well. Willis, Dolan, and Bertrand (1999) have replicated this finding by showing that the most common adherence problem in older adults is "incomplete processing errors," failing to process all the information because of restricted processing resources. Similar findings are seen when adult age differences in medical decision making are investigated (Park, 1999). When making decisions about complex treatment options, older adults often rely on less information and consider fewer options, even when provided with the same amounts of information given to younger adults (Meyer, Russo, & Talbot, 1995; Zwahr, Park, & Shifren, 1999). Reduced processing resources mean that older adults will often have less processed information available to them when making decisions.

In a recent study of rheumatoid arthritis patients, however, Park et al. (1999) found that although working memory, a good measure of processing resources, and other memory deficits predict nonadherence, the oldest group of arthritis patients, the group with the poorest working memory, were the most adherent. The middle-age group was the most nonadherent, most likely because of their busy lifestyles. Older adults were the most adherent, even with memory factors that put them at greater risk. Medication adherence, therefore, is not just a memory problem, but a matter governed by social and other cognitive factors (Park et al., 1999).

From a human factors perspective, design issues involve conveying the information to the older patient in a manner that ensures easy comprehension and providing reminders for taking medications. Morrow, Leier, and Sheikh (1988) have developed recommendations for adherence designs based on cognitive aging research. Morrow and Leirer (1997) suggested that well-designed instructions could reduce age differences by providing environmental support, either through easily organizable text or the use of simple icons. These format considerations reduce the processing required for comprehension and remembering. Because medication adherence often is a prospective memory task (remembering to take medicine in the future), providing reminders at the time medication is required would make the prospective memory task an event-based task rather than one in which the only cue is the passage of time. Adherence by older adults is more likely because the contextual cue for adherence is more salient, and the reminder provides environmental support. In fact, research shows that age differences in event-based prospective memory tasks are smaller than in time-based prospective tasks (Park, Hertzog, Kidder, Morrell, & Mayhorn, 1997).

Much of the preceding discussion on medication adherence deals with the use of medication instructions. When the deliberate processing requirements of the task are reduced by providing environmental support, age differences are reduced. The same principles used in the home by the patient him- or herself can be applied in the design of medical devices.

Design of Medical Devices

Gardner-Bonneau and Gosbee (1997) in their review of health care and human factors indicate that there is a paucity of research aimed at human factors, aging, and health care. Yet, medical technologies are constantly changing, and there is greater reliance on home health and self-care. These authors go on to propose relevant research-based design principles to be used by human factors experts when designing medical devices to be used by older adults. They suggest that the number of steps needed to operate the device should be small to avoid processing resource requirements and difficult memorization. They also suggest that an attempt should be made to use features common to other familiar devices used by older adults. This is an excellent suggestion, which takes advantage of spared knowledge and semantic memories of older adults. Clearly, usability testing for new devices should determine specific problems faced by older adults using these devices, including memory problems. Czaja (1997) pointed out how clock radios that verbalize reminders could greatly aid the prospective memories of older adults, turning time-based tasks into event-based tasks.

SUMMARY

Table 3.3 provides some examples of intervention guidelines based on the seven conclusions about aging and memory. They are grouped by the three intervention strategies suggested by Baltes and Baltes (1990): selection, optimization, and compensation. The selection strategy suggests that memory skills showing little decline with aging should be emphasized (i.e., skills that rely on existing knowledge and semantic memory and skills based on well-learned and easily remembered procedures). The selection strategy also suggests that the processing resources needed to meet any memory requirement of the task be kept as small as possible to avoid working memory and processing resource limitations.

The optimization strategy is based on the fact that the performance of older adults is rarely at their reserve capacity, and that practice and training can be used

TABLE 3.3

Intervention Strategies and Example of Intervention Guidelines Based on Cognitive Aging Research
(Numbers Represent Entries in Table 3.1)

Intervention Strategy	Example Intervention Guidelines
Selection	• In medical instructions and designing training tasks, take advantage of increased world knowledge, expertise, and semantic memories of older adults. Procedures for operating new devices should be based on familiar procedures already used by older adults. (1, 4, 5, 7) • Use well-integrated (relevant to response), familiar cues and reminders of medical adherence. (1, 2, 5) • In designing instructions or medical devices, keep tasks simple enough to avoid resource processing demands. (6, 7) • If memory is required, rely more on tasks that involve fewer cognitive resources (e.g., cued recall and recognition rather than unguided free recall.) (6, 7)
Optimization	• Use more training with older adults, taking advantage of plasticity of performance on most cognitive tasks. (4, 6, 7) • Provide sufficient practice with task during training and during use period, to optimize performance. (4, 5, 6) • Be careful in making sure that all distracting or irrelevant information is eliminated from the instructions or from the performance of the task. (3)
Compensation	• Create conditions of practice that will increase the extent that remembering will be based on familiarity rather than intentional deliberate processing. (5, 6, 7, 8) • Use more external memory aids to minimize self-initiated explicit recall. (6, 7)

to improve performance. Although more extensive practice or training may be necessary and sensitivity to cognitive aging in the design of instructions will be needed, there is sufficient plasticity in behavior to improve performance. The optimization strategy also suggests that care should be taken to avoid distraction and irrelevant information that could tax the cognitive resources of older adults and reduce the probably of their achieving the desired performance.

Finally, compensation suggests that different ways of achieving the desired performance should be considered. For example, increasing the automaticity of the memory task would cause it to be based more on familiarity and habit and less on deliberate recollection. An obvious compensatory strategy is to rely more on external memory aids (environmental support) instead of relying solely on cognitive recollection.

REFERENCES

Baltes, P. B., & Baltes, M. M. (1990). Psychological perspectives on successful aging: The model of selective optimization with compensation. In P. B. Baltes & M. M. Baltes (Eds.),

Successful aging: Perspectives from the behavioral sciences (pp. 1–34). New York: Cambridge University Press.

Baltes, P. B., & Lindenberger, U. (1997). Emergence of a powerful connection between sensory and cognitive functions across the adult life span: A new window to the study of cognitive aging? *Psychology and Aging, 12,* 12–21.

Clancy, S. M., & Hoyer, W. J. (1994). Age and skill in visual search. *Developmental Psychology, 30,* 545–552.

Craik, F. I. M., & Anderson, N. (1999). Applying cognitive research to problems of aging. In D. Gopher & A. Koriat (Eds.), *Attention and Performance XVII. cognitive regulation of performance: Interactions of theory and application.* Cambridge, MA: MIT Press.

Craik, F. I. M., & Byrd, M. (1982). Aging and cognitive deficits: The role of attentional resources. In F. I. IM. Craik & S. Trehub (Eds.), *Aging and cognitive processes* (pp. 191–211).

Craik, F. I. M., & Jennings, J. M. (1992). Human memory. In F. I. M. Craik & T. A. Salthouse (Eds.), *Handbook of aging and cognition* (pp. 384–420). Hillsdale, NJ: Lawrence Erlbaum Associates.

Craik, F. I. M., & McDowd, J. M. (1987). Age differences in recall and recognition. *Journal of Experimental Psychology: Learning, Memory, and Cognition, 13,* 474–479.

Czaja, S. J. (1997). Using technologies to aid the performance of home tasks. In A. D. Fisk, & W. A. Rogers (Eds.). *Handbook of human factors and the older adult.* (Pp. 311–334). San Diego: Academic Press.

Deeg, D. J. H,., Kardaun, J. W. P. F., & Fozard, J. L. (1966). Health, behavior, and aging. In J. E. Birren & K. W. Schaie (Eds.), *Handbook of the psychology of aging,* (Fourth Edition, Pp. 129–149). San Diego: Academic Press.

Duke University Center for the Study of Aging and Human Development. (1978). *Multidimensional Functional Assessment: The OARS methodology.* Durham, NC: Duke University Press.

Earles, J. L. K., Connor, L. T., Smith, A. D., & Park, D. C. (1997). Interrelations of age, self-reported health, speed, and memory. *Psychology and Aging, 12,* 675–683.

Earles, J. L. K., & Salthouse, T. A. (1995). Interrelations of age, health, and speed. *Journal of Gerontology: Psychological Sciences, 50B,* P33–P41.

Earles, J. L., Smith, A. D., & Park, D. C. (1994). Age differences in the effects of facilitating and distracting context on recall. *Aging and Cognition, 1,* 141–151.

Freund, A. M., & Baltes, P. B. (1998). Selection, optimization, and compensation as strategies of life management: correlations with subjective indicators of successful aging. *Psychology and Aging, 13,* 531–543.

Gardner-Bonneau, D., & Gosbee, J. (1997). Health care and rehabilitation. In A. D. Fisk, & W. A. Rogers (Eds.). *Handbook of human factors and the older adult.* (Pp. 231–256). San Diego: Academic Press.

Hasher, L. & Zacks, R. T. (1988). Working memory, comprehension, and aging: A review and a new view. In G. Bower (Ed.), *The psychology of learning and motivation* (Vol. 22, pp. 193–225). New York: Academic Press.

Hay, J. F., & Jacoby, L. L. (1999). Separating habit and recollection in young and older adults: Effects of elaborative processing and distinctiveness. *Psychology and Aging, 14,* 122–134.

Howard, D. V. (1996). The aging of implicit and explicit memory. In F. Blanchard-Fields & T.M. Hess (Eds.), *Perspectives on cognitive change in adulthood and aging* (pp. 221–254). New York: McGraw-Hill.

Hultsch, D. F., Hammer, M., & Small, B. J. (1993). Age differences in cognitive performance in later life: Relationships to self-reported health and activity life style. *Journal of Gerontology: Psychological Sciences, 48,* P1–P11.

Jacoby, L. L. (1991). A process dissociation framework: Separating automatic from intentional uses of memory. *Journal of Memory and Language, 30*, 513–541.

Jennings, J. M., & Jacoby, L. L. (1997). An opposition procedure for detecting age-related deficits in recollection: Telling effects of repetition. *Psychology and Aging, 12*, 352–361.

Kausler, D. H., & Kleim, D. M. (1978). Age differences in processing relevant versus irrelevant stimuli in multiple item recognition learning. *Journal of Gerontology, 33*, 87–93.

Light, L. L. (1996). Memory and aging. In E. L. Bjork & R. A. Bjork (Eds.), *Memory* (pp. 443–490). San Diego: Academic Press.

Light, L. L., & Singh, A. (1987). Implicit and explicit memory in young and older adults. *Journal of Experimental Psychology: Learning, Memory, & Cognition, 13*, 531–541.

Luszcz, M.A., Bryan, J., & Kent, P. (1997). Predicting episodic memory performance of very old men and women: Contributions from age, depression, activity, cognitive ability, and speed. *Psychology and Aging, 12*, 340–351.

Meyer, B. J. F., Russo, C., & Talbot, A. (1995). Discourse comprehension and problem solving: Decisions about the treatment of breast cancer by women across the life span. *Psychology and Aging, 10*, 84–103.

Morrell, R. W., Park, D. C., Kidder, D. P., & Martin, M. (1997). Adherence to anithypertensive medications across the life span. *The Gerontologist, 37*, 609–617.

Morrow, D., & Leirer, V. O. (1997). Designing medication instructions for older adults. In D.C. Park, R.W. Morrell, & K. Shifren (Eds.), *Processing of medical information in aging patients.* (Pp. 249–266). Mahwah, NJ: Lawrence Erlbaum Associates.

Morrow, D., Leirer, V. O., & Skeikh, J. (1988). Adherence and medication instructions: Review and recommendations. *Journal of the American Geriatrics Society, 36*, 1147–1160.

Park, D. C. (1992). Applied cognitive aging research. In F. I. M. Craik, & T. A. Salthouse (Eds.), *Handbook of aging and cognition* (pp. 449–494). Hillsdale, NJ: Lawrence Erlbaum Associates.

Park, D. C. (1999). Aging and controlled and automatic processing. In D. C. Park, R. W. Morrell, & K. Shifren (Eds.). *Processing of medical information in aging patients.* (pp. 3–22). Mahwah, NJ: Lawrence Erlbaum Associates.

Park, D. C., Hertzog, C., Kidder, D., Morrell, R., & Mayhorn, C. (1997). The effect of age on event-based and time-based prospective memory. *Psychology and Aging, 12*, 314–327.

Park, D. C., Hertzog, C., Leventhal, H., Morrell, R. W., Leventhal, E., Birchmore, D., Martin, M., & Bennett, J. (1997). Medication adherence in rheumatoid arthritis patients: Older is wiser. *Journal of the American Geriatrics Society, 47*, 172–183.

Park, D. C., & Jones, T. R. (1997). Medication adherence and aging. In A. D. Fisk & W. A. Rogers (Eds.), *Handbook of human factors and the older adult.* (pp. 257–287). San Diego, CA: Academic Press.

Park, D. C., Morrell, R. W., Frieske, D., & Kincaid, D. (1992). Medication adherence behaviors in older adults: Efects of external cognitive supports. *Psychology and Aging, 7*, 252–256.

Park, D. C., Smith, A. D., Dudley, W. N., & Lafronza, V. N. (1989). The effects of age and a divided attention task presented at encoding and retrieval on memory. *Journal of Experimental Psychology: Learning, Memory, and Cognition, 15*, 1185–1191.

Park, D. C., Smith, A. D., Lautenschlager, G., Earles, J., Frieske, D., Zwahr, M., & Gaines, C. (1996). Mediators of long-term memory performance across the life span. *Psychology and Aging, 11*, 621–637.

Parkin, A. J., & Walter, B. M. (1992). Recollective experience, normal aging, and frontal dysfunction. *Psychology and Aging, 7*, 290–298.

Rogers, W. A. (1992). Age differences in visual search: Target and distractor learning. *Psychology and Aging, 7*, 526–535.

Rogers, W. A., and Fisk, A. D. (2000). Human factors, applied cognition, and aging. In F. I. M. Craik & T. A. Salthouse (Eds.), *The handbook of aging and cognition* (2nd edition, pp. 559–591). Mahwah, NJ: Lawrence Erlbaum Associates.

Schaie, K. W. (1996). Intellectural development in adulthood. In J. E. Birren & K. W. Schaie (Eds.), *Handbook of the psychology of aging*, (4th ed. pp. 266–286). San Diego: Academic Press.

Smith, A. D. (1996). Memory. In J. E. Birren & K. W. Schaie (Eds.), *Handbook of the psychology of aging* (4th ed, pp. 236–250). San Diego: Academic Press.

Smith, A. D., Park, D. C., & Whiting, W. (1998). *Differential effects of facilitating and distracting context on age differences in cued recall.* Paper presented to the Cognitive Aging Conference (Atlanta, April).

Stankov, L., & Anstey, K. (1997). Health and cognitive ageing: The emerging role of sensorimotor abilities. *Australian Journal on Ageing, 16,* 34–39.

Tulving, E., & Schacter, D.L. (1990). Priming and human memory systems. *Science, 247,* 301–306.

Willis, S. L., Dolan, Dolan, M. W., & Bertrand, R. M. (1999). Problem solving on health-related tasks of daily living. In D. C. Park, R. W. Morrell, & Shifren, K. (Eds.), *Processing of medical information in aging patients* (pp. 199–219). Mahwah, NJ: Lawrence Erlbaum Associates.

Yates, J. F., & Patalano, A. L. (1999). Decision making and aging. In D. C. Park, R. W. Morrell, & K. Shifren (Eds.), *Processing of medical information in aging patients* (pp. 31–54). Mahwah, NJ: Lawrence Erlbaum Associates.

Yesavage, J. A., Lapp, D., & Sheikh, J. I. (1989). Mjnemonics as modified for use by the elderly. In L. W. Poon, D. C. Rubin, & B. A. Wilson (Eds.), *Everyday cognition in adulthood and late life* (pp. 598–611). New York: Cambridge University Press.

Zacks, R. T., Hasher, L., & Li, K. Z. H. (2000). Human memory. In F.I.M. Craik, & T. A. Salthouse (Eds.), *Handbook of aging and cognition* (2nd ed., pp. 293–357). Mahwah, NJ: Lawrence Erlbaum Associates.

Zwahr, M. D. (1999). Cognitive processes and medical decisions. In D. C. Park, R. W. Morrell, & K. Shifren (Eds.), *Processing of medical information in aging patients* (pp. 55–68). Mahwah, NJ: Lawrence Erlbaum Associates.

4

Cognitive-Linguistic Aging: Considerations for Home Health Care Environments

Constance Dean Qualls
The Pennsylvania State University

Joyce L. Harris
The University of Memphis

Wendy A. Rogers
Georgia Institute of Technology

Living longer, and living longer with disability, older adults are the fastest growing population in the United States (Hobbs & Damon, 1996). Despite unprecedented longevity and survivability, their access to public and private health care has become increasingly restricted, in part, because of dramatic changes in health care costs. The combination of greater need and fewer resources has ushered in the use of a broader spectrum of health care environments, including subacute and long-term care, adult day care, and, increasingly, home health care. Ongoing economic changes in health care finance have resulted in the reclamation of the home environment as the locus of choice for long-term care provision (Bogner, 1999) and has spawned a new health care industry, represented by home health agencies.

Home health care, in today's parlance, refers to the at-home rendering of health-related services by paid skilled health care professionals. Kane (1999), however, does not limit her definition to skilled personnel, broadly defining home care as "any health-related care or help from paid personnel received by people who live at home and require the help because of their health and functional status" (p. 300).

47

Formal home-based health care services for individuals with illness or disability may range from skilled nursing and rehabilitation services (e.g., speech-language pathology, physical therapy, and occupational therapy) to personal care assistance with bathing, dressing, feeding, and housekeeping (Feldman, 1999). The physical, cognitive, communicative, and affective status of the patient is key in determining the level of intervention.

Home health care environments also can be defined as those in which unskilled individuals (i.e., family members, friends, nursing aides, or domestic workers) provide care (see Schulz, Czaja, and Belle, chap. 8, this volume). So-called informal home health care often is provided by older individuals (i.e., a spouse) whose own physical and cognitive functioning may be compromised by medication or age-related cognitive decline (Bogner, 1999). Perhaps a more frequently overlooked home health care environment is one in which there is no observable caregiver. That is, the older adult, deemed functionally independent, lives alone and provides self-care.

At-home health care, however defined, presents special challenges for older adults, including those who are considered functionally independent. Shorter hospital stays dictate that people return to their homes "quicker and sicker" than previously would have been thought medically advisable. Thus, the current standard of care, while believed to be more economically acceptable, can lead to longer periods of convalescence and dependency, especially for older adults.

Successful negotiation of home health care environments requires that individuals possess intact auditory-verbal and written language skills. Auditory-verbal comprehension, the ability to understand spoken language, is requisite for understanding medical information and complying with the prescribed interventions. Examples include fully understanding one's medical condition and current health status, patient rights, and privileges (e.g., consent information), appointment information (including dates, times, locations, pre-requisites, and follow-up directives), and discharge instructions. Appropriate use of prosthetic and other ancillary devices (e.g., insulin pumps, blood pressure monitors) necessitate the ability to learn and use these devices effectively. Such activities are accomplished through auditory-verbal provider–patient interactions that typically are supplemented by written materials.

Health care literature such as pamphlets, brochures, and newsletters are useful aids to comprehension. Written materials, when appropriately prepared, provide visual cueing and redundancy, and thus can reduce memory load and enhance comprehension, especially for older adults. Taking into account the ubiquitous nature of published literature on matters of health care, including and especially that on the Internet, comprehension of written materials is vital for independent functioning within home health care environments. Older adults are expected to be able to understand and use following types of health care literature:

- Medication labels containing information regarding dosage, frequency of intake, and medication interactions.

- Warning labels on over-the-counter medications.

- Information-laden pamphlets included with prescription medications describing generic equivalents and side effects.

- Postoperative and discharge instructions.

- User manuals for home health care technology such as blood glucose meters, infusion pumps, and blood pressure monitors.

- Nutrition labels on foods and food supplements.

- Instructions for assistive devices and medical equipment such as hoists and wheelchairs.

- Installation instructions for environmental accommodations such as grab bars and wheelchair ramps.

- Medicolegal documents such as insurance forms, living wills, and consent forms.

Traditional face-to-face auditory-verbal communication and consumer literature are being augmented rapidly by new communication modalities, for example, automated telephone, and other computer-based technologies such as telemedicine and the World Wide Web (Morrow & Leirer; chap. 10, Stronge, Walker, & Rogers, chap. 14, and Whitten, chap. 7, this volume). Because of age-related changes in language and cognition, some older adults also experience difficulty obtaining and using information from these sources.

Considerable evidence suggests that older adults show a decline in their language comprehension as they get older (Cohen, 1979; Nicholas, Connor, Obler, & Albert, 1998; Obler, Au, & Albert, 1995). The degree of decline in language comprehension is related to the amount of stress placed on the cognitive-linguistic system, for example, increased cognitive demands or removal of contextual cues that facilitate comprehension. Age-related decline in discourse comprehension (i.e., understanding of longer, text-based material) is evident when time constraints are imposed, or when inferencing is required (Cannito, Hayashi, & Ulatowska, 1988; Shadden, 1997).

In home health care environments, adequate language comprehension skills become especially important for older adults considering the nature of the communication process. Auditory-verbal interactions most often are rapidly-paced and contain a lot of information, which is sometimes unclear, incomplete, or ambiguous. Written materials also present the older reader with certain challenges. For example, readability (i.e., grade level, number of passives, vocabulary) and typography (i.e., type size, spacing, type style) issues may interfere with comprehension. For older adults, communication can be further compromised by sensory deficits (i.e., hearing, visual impairments), cognitive declines (i.e., reduced speed of processing, memory, and attentional deficits), and language impairments (i.e., word finding difficulties, misinterpretation of nonliteral language).

Because of the ever-changing health care system in the United States, health care and human factors researchers have begun to investigate ways to improve older adults' ability to function effectively and easily in the home health care environment. Paramount to this charge is gaining insight about the cognitive and language (cognitive-linguistic) abilities necessary to support activities for living at home with medical conditions. Little is known in this regard. Therefore, the goals for this chapter are to: (a) describe the general nature of the cognitive-linguistic changes that take place in normal aging, with particular emphasis on comprehension; (b) detail the cognitive-linguistic alterations that occur with stroke and dementia; (c) discuss cognitive-linguistic aging from the standpoint of health care information, processing and use; and, (d) outline a preliminary research agenda for health care professionals and human factors researchers regarding aspects of cognitive-linguistic aging in home health care environments.

COGNITIVE-LINGUISTIC CHANGES IN NORMAL AGING

Theoretical Considerations

Language is used to communicate thoughts and ideas, and language-use is one way in which thoughts and ideas are conveyed. Domains of language include phonology, morphology, syntax, semantics, and pragmatics. *Phonology* refers to the sounds of linguistic symbols, whereas *morphology* deals with the form and structure of linguistic units. *Syntax* is concerned with the grammatic ordering of linguistic symbols, whereas *semantics* deals with the relation between symbols and real or imagined objects, events, and ideas. *Pragmatics* refers to the relations between symbols, signs, and their users, and therefore encompasses the communicative goals of the speaker. *Cognition* is broadly defined as the act of and the mental processes involved in perceiving, remembering, and thinking (Ashcraft, 1989). The nature of the relation between language and cognition continues to challenge researchers with respect to developing theories of language processing in aging.

Theorists do agree that there is not a one-to-one relation between language and cognition. Some aspects of cognition are immune to linguistic expression (e.g., activities involving procedural knowledge) and language contains many arbitrary conventions and constraints that appear to be purely linguistic in nature (e.g., shifting of word categories such as nouns and pronouns with no corresponding shifts in meaning) (Kemper, 1992). Nevertheless, both cognitive and linguistic mechanisms are involved with perception of speech and language. That is, language comprehension and production (e.g., remembering, understanding, paraphrasing, producing, evaluating) occur within the constraints of the information processing system, specifically memory and attention (Carroll, 1986).

Research results show age differences in performance between younger and older adults on such cognitive tasks as reasoning, spatial ability, inhibition, speed of processing, and memory (Baddeley, 1986; Craik & Jennings, 1992; Salthouse, 1992; Zacks & Hasher, 1994). Other studies have found age differences in a variety of linguistic domains, including syntax and semantics (Kemper, 1992; Obler, Fein, Nicholas, & Albert, 1991). Age differences in lexical retrieval and discourse also have been observed (Au et al., 1995; Cannito et al., 1988). The findings from these studies suggest that, whereas age differences do in fact exist, the magnitude and the underlying reasons for these differences are not always clear. Nonetheless, older adults show age-related declines in certain cognitive abilities, and some of these changes, for example, inhibitory inefficiency, reduced processing speed, and memory changes (i.e., deficits in working memory), have been shown to have a detrimental effect on their language abilities. Cognitive-linguistic aging is concerned with age-related performance changes in various measures of linguistic functioning associated with age-related cognitive changes.

Working memory, for example, plays a major role in language comprehension, especially for complex tasks such as understanding syntactically complex sentences and tasks requiring inference (Cohen, 1979; Kemper, 1992). Working memory is a theoretical construct used to describe the short-term memory system involved in temporary processing and storage of information (Baddeley & Hitch, 1974). A limited-capacity processor, working memory supports a number of cognitive activities, including reasoning, learning, mental calculation, and language comprehension (Baddeley, 1986; Baddeley & Hitch, 1974; Gathercole & Baddeley, 1993; Kausler, 1994). Working memory restrictions and generalized cognitive resource reductions have been proposed to explain age-related changes in language and communicative behaviors (Shadden, 1997; Ulatowska & Chapman, 1991).

Although most research suggests age-related decline in cognitive functioning, some research findings indicate that cognitive decline is not a necessary consequence of age. Two theoretically opposed views have emerged in this regard. The *decrementalist view* suggests that cognitive decline is inevitable, whereas the *continued-potential view* suggests that some cognitive skills actually increase with age (see Lemme, 1999 for a complete discussion). According to Perlmutter (1988), some cognitive skills may even emerge in later adulthood. Contemporary aging researchers now agree that both growth and decline occur in adult development, and that the assumption of inevitable cognitive decline in all older adults is false at best (Lemme, 1999). Because individuals in home health care environments often have chronic and disabling health conditions, many demonstrate lower levels of cognitive functioning, which often have an impact on their language abilities. Individual differences must be considered, however, particularly because the cognitive abilities maintained into old age may support language functions.

Language Comprehension

Language comprehension is a product of the ability to compute semantic and syntactic relations among successive words, phrases, and sentences for coherence and meaning, and to integrate new information with previously processed information (Daneman & Merikle, 1996). Thus, memory interacts significantly with linguistic processing. In addition to memory, language comprehension is supported by other cognitive processing activities, including semantic activation, registration and analysis of the surface meaning (recall of the exact words and sentence structure), integration of novel information with previous knowledge, and inference of meaning within the context provided (Burke & Yee, 1984; Cohen, 1979). Cross-sectional studies of language comprehension show that changes in some or all of these processes produce age-associated performance deficits for older adults (Cohen, 1979; Wingfield & Stine-Morrow, 2000). Language studies also have found age-related decline in comprehension of nonliteral language such as idioms, metaphors, and metonyms (Qualls, 1998; Vogel, Sugar, & Cardillo, 1995; Zelinksi & Hyde, 1996)

In addition to cognition, other factors can affect language processing in older adults. Changes in sensory abilities such as visual perception and hearing as well as changes in affect, motivation, and social status also may play a role. Recent technological advances place even greater demands on linguistic processing of both auditory-verbal and written language, especially for older adults (Tun & Wingfield, 1997). Elaborate and sometimes cumbersome telephone menus, automatic teller machines, quick camera shots and sound bites on television, digital telephone answering machines and other digitized "talking" devices, and computer and Web-based technology all present visual, perceptual, cognitive, and linguistic challenges to many older adults.

The ability to comprehend and use information from language is essential for independent functioning of the individual across the life span and into old age (Tun & Wingfield, 1997). Language processing requirements in home health care environments presume a specified level of independence in remembering, comprehending, and appropriately acting on information. Instructions for diet and medications, self-care activities, personal appointments, and daily communicative interactions depend on such skills (Tun & Wingfield, 1997). However, older adults demonstrate changes in working memory, speed of processing, and attention, and for some older adults, these changes can promote the use of alternative processing styles and strategies, which subsequently produce changes in language function (Kemper, 1992). For older adults, memory deficits can lead to word retrieval failures (Au et al., 1995). Consequently, word retrieval difficulties can lead to inadequate language output, which ultimately can prevent communication of the intended verbal message. This has implications for the ability of older adults to access much-needed health care services in the home health environment.

Preserved Language Abilities in Aging

Not all older adults show linguistic processing deficits that interfere with everyday functioning, possibly because of mediating factors. For example, although findings are mixed across studies, considerable evidence indicates that older adults show a decline in some language abilities, whereas certain language abilities are relatively unaffected by age. Preserved language abilities include highly overlearned language (e.g., greetings, days of the week), word recognition, basic lexical and semantic skills (e.g., retention of underlying semantic meanings), phonology, and syntax (Burke & Yee, 1984; Cerella & Fozard, 1984; Shadden, 1997).

It is possible that for some older adults, preserved language abilities may facilitate language comprehension. Several theoretical explanations for linguistic compensations have been proposed. The *selective optimization with compensation* model for successful aging (Baltes, Dittmann-Kohli, & Dixon, 1984; see also Smith, this volume) suggests that the individual maximizes strengths (i.e., enriched vocabulary) while developing compensations for weaknesses (e.g., the use of contextual cues to facilitate comprehension). Salthouse (1991) proposed a three-stage model of successful aging that includes *accommodation* (redistribution of cognitive effort for more efficient use of resources), *compensation* (adjustment of strategies to increase efficiency), and *remediation* (cognitive intervention for restoration of cognitive abilities). Specific to language processing, Tun and Wingfield (1997) suggested that mediation of age-related decline may be a function of the inherent *linguistic structure* (phonetic, syntactic, semantic, and pragmatic constraints), *prosodic facilitation* (use of pitch contour, stress, and timing), and *predictability* (listener-expectation of the intended message based on the given context). It may be that for older adults who use linguistic compensation during comprehension tasks, age differences do not emerge. Therefore, linguistic compensation may account for some of the individual differences seen in older adults.

In addition to preserved language abilities, other factors such as level and quality of education, socioeconomic status, health status, and task requirements can mediate age differences in performance of language tasks. Cohen's (1979) work on language comprehension in aging suggests that older adults with high levels of education and relatively good health have better language processing skills than those who do not fall into these categories. Harris, Rogers, and Qualls (1998) also demonstrated that older adults with high levels of education and health status performed as well as their younger adult counterparts in verifying statements about previously read expository, narrative, and procedural passages. In a follow-up study, Qualls, Harris, and Rogers (2000) replicated their original findings on the identical recognition task, using participants with an overall mean level of education significantly lower than that of the adults in the first study. However, on a comparable recall task, also administered in the follow-up study, age differences did emerge. Findings from these extant studies suggest that observed age differences in language comprehension may depend on the education level of the participants, as well as the task requirements. It may be that individuals with higher levels of education have an increased ability to compensate for reduced or declining linguistic capacity. In other words, they may have a wider repertoire of facilitative strategies at their disposal than individuals with lower levels of education.

Strategy to Enhance Comprehension in Aging

For some older adults, strategy can enhance comprehension of auditory-verbal and written information. In their work investigating the effect of self-selected memory strategies on comprehension, Harris and Qualls (2000) found that older adults who used elaborated memory-enhancing strategies to complete a verbal working memory task performed significantly better in their reading comprehension than those who used maintenance rehearsal.

To maximize reading comprehension and retention in older adults, Meyer, Talbot, Stubblefield, and Poon (1998) suggested presenting coherent texts containing explicit signals that highlight the main ideas and relations contained in the text, especially when the topic of the text is uninteresting to the older adult. Lending support for Kintsch's (1980) model of text comprehension, the findings of Meyer et al. (1998) suggest that signaling increases the structural cohesion of text, which in turn "reduces processing resource demands and generates text-based situational characteristics that may support the development of cognitive interest" (p. 768). Harris, Rogers, and Qualls (1998) found that repeated presentation of texts enhanced comprehension, especially for older adults. Moreover, regarding strategy use in home health care environments, Harris and Qualls (2000) stated:

The need for additional demonstrations of strategic processing's effectiveness continues as a compelling charge to researchers who are engaged in communication in aging research, as the need for functional linguistic processing skills across the life span must coexist with, and supercede, increasing threats to cognitive-linguistic resources in old age. (p. 12)

COGNITIVE-LINGUISTIC CHANGES RESULTING FROM STROKE AND DEMENTIA

The dramatic demographic shift in this country has considerable implications for research scientists investigating both normal and pathologic aging. Because people are living longer, it is expected that there will be an increased prevalence of communication disorders among older adults, as implied by cerebrovascular disease and stroke statistics. In addition to people with stroke, the number of individuals with dementia (intellectual decline) will grow at alarming rates. Thus, many elderly individuals will experience impoverished cognitive and language processing skills. Language impairments resulting from stroke and dementia can manifest in one or more components of language including phonology, morphology, syntax, semantics, and pragmatics, and may coexist with motor and sensory deficits such as speech and hearing difficulties. Both stroke and dementia imply that there will be a continuing need for home health care.

Language Changes Resulting From Stroke

The effects of normal aging on language comprehension can be exacerbated significantly by brain damage. The National Stroke Association (1999) reported the following facts about stroke:

- Stroke is the third leading cause of death after heart disease and cancer.

- Stroke incidence is approximately 730,000 new cases each year.

- Approximately 4 million people have experienced a stroke.

- Approximately four in every five families are affected by stroke.

- Stroke has a 30% mortality rate.

- Of all individuals with stroke, 20% to 30% are severely and permanently disabled.

- Age factors significantly in stroke.

- An individual's chances of having a stroke doubles for every decade after the age of 55 years.

- Two thirds of strokes occur in individuals older than 65 years.

- There is a greater chance of nonwhites (especially African Americans) having a stroke at a younger age than whites.

Stroke, the leading cause of acquired language disorders in adults, has been identified as a major and increasing factor in late-life dementia, especially for individuals older than 80 years. Between 80,000 and 100,000 new cases of aphasia result from stroke each year. Aphasia is a language disorder characterized by a disturbance of comprehension and language formulation caused by damage to specific areas of the brain (Damasio, 1992). Individual differences in aphasia deficits are seen across patients based primarily on differences in site of lesion, severity of damage, premorbid functioning, and associated medical conditions. For this reason, aphasia can compromise single or multiple aspects of language such as phonology, morphology, and syntax (Damasio, 1992). Largely depending on the site of lesion and level of severity, language difficulties can be seen in comprehension, verbal production, reading, writing, gestures, and nonverbal behaviors (Vinson, 1999).

The increased prevalence of stroke in adults 65 years of age and older has considerable implications for the home health care environment because many more people now than ever before are living at home with speech, language, and cognitive deficits after experiencing a stroke. However, because aphasia represents a disruption in linguistic performance and not communicative competence (McNeil, 1982), retraining and rehabilitation are viable options for most individuals with aphasia after a stroke. Speech-language pathologists take the leading role in aphasia rehabilitation, providing services in the home of many individuals living with stroke.

Language Changes From Dementia

Commonly found in individuals older than 65 years, *dementia* is a term used to describe the progressive loss of cognitive and intellectual function. Alzheimer's dementia (AD), constituting 50% to 65% of all dementias, has been diagnosed in approximately 4 million Americans, and the numbers are expected to reach approximately 14 million by the middle of the next century (Alzheimer's Disease and Related Disorders Association, 2000). Characterized by the insidious onset of memory loss and decline in thinking, understanding, reasoning, and decision making, Alzheimer's disease occurs in 1 of 10 persons older than 65 years, and approximately half of the persons older than 85 years.

Normal age-associated language changes and those that occur during the early stages of AD are difficult to differentiate (Chapman, Ulatowska, King, Johnson, & McIntire, 1995) because of overlapping behaviors. For example, in their AD patients, Bayles (1982) and Nicholas, Obler, Albert, and Helm-Estabrooks (1985) reported word-finding problems with phrases that were circumlocutory, indefinite, and lacking in content. *Circumlocution* occurs when a person talks around a word or thought. Often, there is the addition of many words that do not support the intended message (e.g., "that thing that you sit on, near the desk, oh, I don't know, there … : rather than saying "chair"). The most pronounced language deficit in AD is semantic in nature, with grammatical and syntactic processes relatively intact. Consequently, individuals with AD demonstrate impaired naming, paraphasias (substitution of letter sounds and whole words that alter or interrupt the intended meaning of the verbal production), and neologisms (nonsensical word substitutions) in addition to word retrieval deficits (Kempler, 1995; Vinson, 1999).

Recent research findings show that individuals with AD appear able to process relatively complex sentences as they hear them, suggesting that sentence processing deficits in these individuals are likely because of memory difficulties (Kempler, Almor, & MacDonald, 1998). Findings from another study on discourse in AD show that individuals with mild AD have more difficulty integrating the stimulus information with stored knowledge, producing fewer and more partial narratives (i.e., fragments and disruptive responses) when describing an action picture (Chapman et al., 1995).

HEALTH CARE INFORMATION PROCESSING

Literacy in Aging

Clearly, verbal and written language comprehension is a major component of health maintenance. Consequently, literacy is intimately linked to language processing, so low levels of literacy can disadvantage older adults in home health care environments. Literacy, in general, denotes the ability to read, write, calculate, and solve problems, whereas functional literacy is context specific. That is, although an individual may be considered literate in the general sense, when placed in a specific and unfamiliar setting such as the health care environment, the person may lack the ability to function as well. Health literacy is the ability to perform the basic reading and numerical tasks required to function in the health care environment (Ad Hoc Committee on Health Literacy for the Council on Scientific Affairs, American Medical Association [AMA], 1999). Reading and understanding discharge instructions, medication labels, patient education materials, and consent forms are examples of functional health literacy behaviors.

According to the 1992 National Adult Literacy Survey (National Center for Education Statistics, 1993), approximately 44 million American adults are functionally illiterate, with 66% of the individuals in the lowest literacy groups 65 years of age and older. Considering that 21% to 23% of American adults function at the lowest level of literacy, it not surprising that health care environments are experiencing the effects of this phenomenon, especially when the least-literate individuals tend to bear the greatest health burden (AMA, 1999).

Health Literacy in the Elderly: Research Findings

To determine the prevalence of low functional health literacy among older adults, Gazmararian et al. (1999) interviewed 3,260 new Medicare enrollees in a national managed care organization. Participants included both English- and Spanish-speaking individuals from four regions of the country who were 65 years of age or older. The survey was administered orally, and consisted of two parts: (a) questions to determine demographics, self-rated health status, physical functioning, chronic conditions, health care use, mental health, cognitive impairment, social support, and health behaviors; and (b) the Short Test of Functional Health Literacy in Adults (S-TOFHLA; Gazimararian et al., 1999). Their results showed that more than one third of the individuals surveyed had inadequate or marginal health literacy.

Gazmararian et al.'s (1999) results also showed regional and cultural differences in health literacy. For example, lower levels of literacy were more prevalent in Ohio than in Texas and Florida. Also, blacks (11.8% of the sample) and Hispanics (11.2% of the sample) showed lower levels of literacy than whites. A remarkable finding was that adults ages 85 years and older demonstrated higher levels of inadequate or marginal health literacy even after adjustment for education and cognitive impairment, suggesting a strong relation between advancing age and lower levels of literacy.

Limited literacy skills present a barrier to receiving high-quality health care services, especially for older adults. Williams et al. (1995) investigated health literacy in 2,659 predominantly indigent and minority patients in two urban public hospitals using the Test of Functional Health Literacy in Adults (TOFHLA). Their results showed that a high proportion of the patients were unable to read and understand written basic medical instructions, including comprehension of directions for taking medications (42%), understanding of information regarding their next appointment (26%), and understanding of the informed consent document (60%). Among the elderly, 81% of the English-speaking patients and 83% of the Spanish-speaking patients demonstrated inadequate or marginal functional health literacy. Williams et al. (1995) concluded that many patients seen in acute care facilities do not possess adequate health literacy skills to negotiate the health care system and obtain access to high-quality health care.

Inadequate health literacy is correlated with several factors, although it is yet unclear which factors may be more or less related. Nevertheless, low levels of education, socioeconomic status, group membership, and advanced age have been cited as indicators of low health literacy (Gazimararian et al., 1999). Patients with inadequate health literacy experience multifaceted communication difficulties that often affect health outcome adversely. They report worse health status and less understanding about their medical conditions and treatment (AMA, 1999). Consequently, many individuals with low levels of health literacy are at greater risk for frequent hospitalization than individuals with adequate literacy skills. For these individuals, at-home health behaviors may be less than adequate to meet their health care needs.

Cognitive-Linguistic Processing Requirements for Home Health Care Management

Because literacy and language processing are intimately linked, many of the activities carried out in home health care environments require prerequisite health literacy skills. In the following sections, the cognitive-linguistic requirements are discussed, as well as the impact of cognitive-linguistic aging on medication information processing, in-home medical equipment use and maintenance, and warning labels. Cognitive-linguistic processing demands are pervasive for these activities in home health care environments.

Medication Information Processing

Although adults older than 65 years account for less than 15% of the U. S. population, they consume a little more than 30% to 50% of all prescription drugs (Park, Morrell, Frieske, & Kincaid, 1992) and approximately 40% of nonprescription medicines. Successful use of medication requires the ability to read, remember, understand, problem solve, and make decisions for appropriate compliance. Evidence suggests that written instructions accompanied by verbal consultation with the provider increases patient comprehension of medication instructions (German, Klein, McPhee, & Smith, 1982; Morris & Halperin, 1979). For example, De Tullio et al. (1986) investigated whether the patient medication instructions (PMI) alone or with provider consultation would enhance patient knowledge. Their results indicate that patients who received PMI information from a provider, with or without a verbal consultation, demonstrate higher levels of knowledge than those who received the written PMI from pharmacists. On the basis of these findings, it may be that directed attention to particular aspects of the written information by the health care provider helps the patient to assign saliency. Once relative importance has been attached to particular information, it may be that the patient will use strategies (e.g., rereading, rehearsal, or questioning) to facilitate comprehension.

In-Home Medical Equipment Use

Following instructions to assemble and use medical equipment presents both physical and cognitive challenges to adults in home health care environments. Information-processing requirements for the appropriate use of medical equipment is increasingly becoming of interest to researchers investigating aging (Bogner, 1999), and can include such activities as reading, understanding, interpreting, and inferencing. In this regard, it is important to consider the caregiver as well as the patient, especially when the caregiver is an age peer.

Often, training in the use of medical equipment is done in a stressful environment, for example, just before the hustle and bustle of the patient's discharge, or in a clinical setting as opposed to in the patient's home (see Gardner-Bonneau, chap. 12, this volume). Consequently, the patient and caregiver confront a great deal of emotional stress concerning whether they will remember what to do, when, and how. Training may consist of both verbal and written procedures. It may be that equipment use, in general, may be intimidating, especially when the patient or caregiver has low literacy skills, or little or no experience with such activities.

Bogner (1999) suggested that a host of factors need to be considered regarding home medical equipment use and maintenance (see also Gardner-Bonneau, chap. 12, this volume). Such factors include training; the multitask nature of equipment use; the impact of medication, illness, and aging on cognition; and sensory functioning in caregivers. To compensate for information-processing problems and enhance comprehension, Bogner (1999) suggested that written instructions accompanying medical equipment should accommodate visual deficits by using large font sizes, simplified sentence structure, nontechnical terminology, and a supporting videotaped recording. However, it is critical that the video be well developed because manufacturer-produced videos may not be sufficient for training (Rogers, Mykityshyn, Campbell, & Fisk, 2001).

Warning Labels

Comprehending and adhering to warning labels is of critical importance to the older adult's safe and effective use of prescription and nonprescription drugs, medical equipment, and other products and equipment for daily activities. However, "because of changes in acuity and sensitivity, older adults may have a reduced likelihood of noticing and encoding warnings; because of changes in working memory capacity, older adults may have a reduced likelihood of comprehending warning information and complying with it" (Rousseau, Lamson, & Rogers, 1998, p. 657). Rousseau et al. (1998) suggested that warning labels should be designed to compensate for age-related changes in perception and cognition, suggesting that color, contrast, font and type size, phrasing, and the display of symbols can be adjusted to improve warning processing for older adults.

HEALTH CARE INFORMATION USE

Americans are more health conscious now than ever before, largely because of the abundant information available on health and healthy living. Wellness information access has been a national effort for well over two decades, with the first set of national targets published in 1979. *Healthy People 2000* (released in 1990) provided an agenda to increase the years of healthy living, reduce health disparities in diverse populations, and achieve access to preventive health services. These objectives were targeted primarily through information access at the national, state, and local levels.

Many individuals actively seek information to become more knowledgeable about various health conditions and learn how to modify their lifestyle to prevent or accommodate such conditions, whereas others are passively exposed. Daily media reports about health matters provide up-to-the-minute medical breakthroughs and healthy living tips. Cable networks, books, and other publications dedicated to health are abundant. Access to health information can be obtained readily in libraries and on the World Wide Web. Brochures, pamphlets, flyers, and booklets covering every imaginable disease, disorder, and medical or psychiatric condition are readily available either on display or on request in drug stores and physicians' offices.

Government agencies (e.g., National Institutes of Health–National Institute on Aging) and private foundations (e.g., Alzheimer's Association, American Association of Retired Persons) provide free-flowing informational resources through public service announcements and written materials. Professional organizations, (e.g., the National Aphasia Association, the American Psychological Association, the American Speech-Language-Hearing Association, and the Gerontological Society of America) also provide an abundance of public service information through the World Wide Web and written materials, as well as through their associates. To foster patient–physician relationships, the *Journal of the American Medical Association* (JAMA) launched a Patient Page to provide a means for more effectively communicating with patients (Glass, Molter, & Hwang, 1998). Availability of resources, however, does not ensure adequate information processing and use, especially for older adults with language and cognitive difficulties. In fact, the abundance of information may overwhelm individuals with low literacy.

Nevertheless, printed materials are both a cost-effective and time-efficient means of communicating health messages and the use of such materials for patient education has been an integral component of patient teaching (Wilson, 1996). Health care literature is the primary source for public dissemination of information on health-related issues. Yet, much of this literature's readability continues to be questionable. For example, many flyers, pamphlets, brochures, and booklets are written at a grade level that exceeds the abilities of consumers. The standard reading grade level for written information, such as that found in daily newspapers, is sixth grade (McLaughlin, 1969).

Readability analysis of 15 health care brochures, including 10 from the American Heart Association (AHA) Brain Attack series and 5 from the American Federation of Aging Research (AFAR) Dementia Guidelines for Families series, showed that the literature was written well above a sixth-grade level (Qualls, 2000). Figure 4.1 shows the results of the analysis according to the computer-based Flesch-Kincaid Grade Level Scale. The AHA Brain Attack brochures were written at a mean reading grade level of 7.91, and the AFAR Dementia Guidelines were written at a mean reading grade level of 11.24. Rogers et al. (2001) reported a simi-

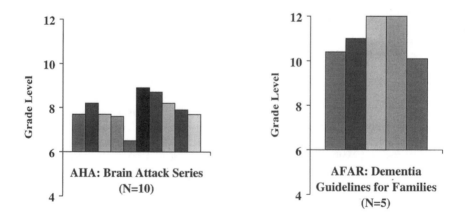

Fig. 4.1   Actual reading grade level of 15 health care brochures. All are written above the recommended sixth-grade level.

lar problem with the readability of the instructions for operating a commonly used blood glucose meter (Figure 4.2). The results of their analysis using the Flesch-Kincaid Grade Level Scale showed that the user manual was written at grade 8 level, the lancing device instructions were written at a grade 6 level, and the test strip instructions were written at grade 9.5 level. Also, the verbal instructions provided with the videotape were written at a grade 7 reading level.

Together, these readability analyses show that existing health care written materials exceed the recommended reading grade level for adults in general. These results present an even greater challenge for some older adults. In addition to reading grade level, the design and layout, font types and sizes, and letter spacing often do not consider visual acuity problems in the elderly. Morrell and Echt (1997) and Hartley (1999) offer recommendations on designing texts and instructions for older users.

DIRECTIONS FOR FUTURE RESEARCH

It would be remiss to end a discussion on the cognitive-linguistic demands of home health care without at least suggesting the beginnings of a research agenda that focuses on determining which theory-based interventions optimize effective, efficient, and economical outcomes for care receivers. Furthermore, the research agenda should focus on the development and evaluation of programs and products designed to ease the burden of formal, informal, and self-care in various home health care environments. The potential for success, the authors believe, rests on a joint scientific effort that uses the expertise of researchers from human factors,

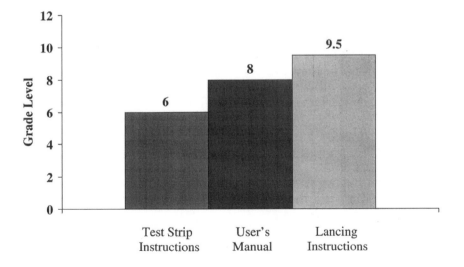

Fig 4.2   Readability of blood glucose meter instructions based on grade level (Rogers et al., 2001).

cognitive and life-span developmental psychology, and communication sciences disciplines. The magnitude of the aging population and the multiple cognitive-linguistic and human factors demands entailed in providing home health care ensures that anything short of a concerted effort will miss the mark. As a starting point, the following suggestions are offered as directions for future research.

Future research should aim

1. To characterize the multiple environments (e.g., group home vs. private residence, apartment vs. single-family dwelling) as well as the multiple permutations of social and psychological circumstances (e.g., care provider and care recipient characteristics, disability and functional level, acceptance of the disability vs. resignation to it) found in each one. A fluid matrix of home care environments should be developed—fluid, because, given individual physical conditions and psychological states, the number of possible human circumstances is nearly infinite. Nevertheless, a well-reasoned matrix representing various home health care environments would underscore the need to develop flexible intervention templates that could be adapted to suit individuals, whose needs may vary from those of others, or even, from one point in time to another.

2. To determine specifically the cognitive-linguistic demands made on care receivers in various home health care environments by constructing a task analysis–driven taxonomy of language processing demands. Again, this is a formidable task, but one thought to be essential for designing effective intervention programs and products for at-home care. One suggestion for organizing a consumer-friendly taxonomy would be to focus on the cognitive-communicative needs associated with activities conducted in specific areas of the home (e.g., the bathroom or the kitchen), thereby associating language needs with care activities. Language comprehension demands must be linked explicitly with specific activities to ease the burden of care for caregivers and to optimize the language processing abilities of care receivers. Thus, a cognitive-linguistic taxonomy of home care activities would further this practical end.

3. To test the efficacy of instruction videotapes for providing health information. In this age of electronic technology, it seems only natural to provide home health care providers with instructional videos, demonstrating the use of medical devices and equipment and the implementation of care routines (e.g., bed-to-chair transfers, swallowing techniques). The use of videotaped instruction should not replace live demonstrations, however. Rather, videotaped instruction should

be used to augment and reinforce instructions provided by health educators. The ready availability and increased affordability of videotape recorders make this a viable supportive strategy for assisting older adults who may not be able to remember the one-time presentation of instructions, or who may want affirmation that they are doing the right thing. Harris et al. (1998) found that repeated reading of text increased language comprehension for older and younger adults. It is hypothesized that the principle of repetition as a strategy for enhancing comprehension also should apply to repeated viewing of instruction videos. Mykityshyn (2000) found that a well-designed video for training users to operate a blood glucose meter was particularly beneficial for older adults relative to a written manual. However, performance was not perfect for either condition, suggesting that future studies should investigate the benefit of video augmentation of written instructions.

4. To test the efficacy of online, interactive distance education for those providing health care in home environments. The information age also should ensure that all older adults have available to them continuing support for their effort to provide at-home health care for themselves and others. Rogers, Cabera, Walker, Gilbert, and Fisk (1996) found that older adults are willing to learn to use computer technology. It may be assumed that the use of computer-based instruction to provide safe and effective health care in the home would be embraced also by older adults. However, the effective delivery of home health care instruction must be informed by the joint research efforts of cognitive psychologists, speech-language pathologists, and human factors experts. Future cohorts of older adults, including the oncoming wave of so-called baby-boomers, will already possess several decades of experience with computer technology as they enter the age of increased risk for providing home health care for themselves or others. However, these individuals will still be susceptible to the perceptual and cognitive declines that can influence technological interactions. Moreover, the technology itself will continue to change.

As stated earlier, these suggestions are merely the beginnings of a rather ambitious research agenda aimed at a better understanding of the cognitive-linguistic and human factors demands inherent to home health care. The aging of society makes the undertaking of this line of research a compelling priority for the early years of the new century because future decades promise continuing increases in medical survivability and longevity. Yet, futurists, many of whom are researchers, must also consider the other side of the coin, which includes the possibility of greater survivability and longevity with disability. Therefore, the need for improved in-home health care looms large.

ACKNOWLEDGMENTS

The first author was supported in part by a grant from the National Institutes of Health-National Institute on Aging, Grant 3 R01 AG14345-03S1, under the auspices of the Language and Aging Brain Lab (Boston University). The second and third authors were supported in part by a grant from the National Institutes of Health–National Institute on Aging, Grant P50 AG1175, under the auspices of the Center for Aging and Cognition: Health, Education, and Training (CACHET). This chapter is based on a presentation made at the CACHET Conference on Human Factors Interventions for the Health Care of Older Adults, February 2000, Destin, Florida.

REFERENCES

Ad Hoc Committee on Health Literacy for the Council on Scientific Affairs, American Medical Association. (1999). Health literacy: Report of the Council on Scientific Affairs. *JAMA, 281*(6), 552–557.

Alzheimer's Disease and Related Disorders Association, Inc. (2000, March). *Alzheimer's disease statistics*. [On-line]. Available: http://www.alz.org. Accessed: 9/30/00.

Ashcraft, M. (1989). *Human memory and cognition*. Glenview, IL: Scott, Foresman.

Au, R., Joung, P., Nicholas, M., Obler, L. K., Kass, R., & Albert, M. L. (1995). Naming ability across the adult life span. *Aging and Cognition, 2*, 300–311.

Baddeley, A. D. (1986). *Working memory*. London: Oxford University Press.

Baddeley, A. D., & Hitch, G. J. (1974). Working memory. In G. Bower (Ed.), *The psychology of learning and motivation*, 8 (pp. 47–90). New York: Academic Press.

Baltes, P. B., Dittmann-Kohli, F., & Dixon, R. A. (1984). New perspectives on the development of intelligence in adulthood: Toward a dual-process conception and a model of selective optimization with compensation. In P.B. Baltes & O.G. Brim, Jr. (Eds.), *Life-span development and behavior* (Vol. 36, pp. 33–76). San Diego, CA: Academic Press.

Bayles, K. A. (1982). Language function in senile dementia. *Brain, 16*, 265–280.

Bogner, M. S. (1999). How do I work this thing? Cognitive issues in home medical equipment use and maintenance. In D. C. Park, R. W. Morrell, & K. Schifren (Eds.), *Processing of medical information in aging patients: Cognitive and human factors perspectives* (pp. 223–232). Mahwah, NJ: Lawrence Erlbaum Associates.

Burke, D. M., & Yee, P. L. (1984). Semantic priming during sentence processing by young and older adults. *Developmental Psychology, 20*, 903–910.

Cannito, M. P., Hayashi, M. M., & Ulatowska, H. K. (1988). Discourse in normal and pathologic aging: Background and assessment strategies. *Seminars in Speech and Language, 9*, 117–134.

Carroll, D. W. (1986). *Psychology of language*. Pacific Grove, CA: Brooks/Cole.

Cerella, J., & Fozard, J. L. (1984). Lexical access and age. *Developmental Psychology, 20*, 235–243.

Chapman, S. B., Ulatowska, H. K., King, K., Johnson, J. K, & McIntire, D. D. (1995). Discourse in early Alzheimer's disease versus normal advance aging. *American Journal of Speech-Language Pathology, 4*, 124–129.

Cohen, G. (1979). Language comprehension in old age. *Cognitive Psychology, 11*, 412–429.

Damasio, A. R. (1992). Aphasia. *New England Journal of Medicine, 33*, 531–539.

Daneman, M., & Merikle, P. M. (1996). Working memory and language comprehension: A meta-analysis. *Psychonomic Bulletin and Review, 3,* 422–433.

De Tullio, P. L., Eraker, S. A., Jepson, C., Becker, M. H., Fujimoto, E., Diaz, C. L., Loveland, R. B., & Strecher, V. J. (1986). Patient medication instruction and provider interactions: Effects on knowledge and attitudes. *Health Education Quarterly, 13*(1), 51–60.

Feldman, P. H. (1999). Doing more for less: Advancing the conceptual underpinnings of home-based care. *Journal of Aging and Health, 3*(11), 261–276.

Gathercole, S. E., & Baddeley, A. D. (1993). *Working memory and language.* Hove, UK: Lawrence Erlbaum Associates, Ltd. Publishers.

Gazmararian, J. A., Baker, D. W., Williams, M. V., Parker, R. M., Scott, T. L., Green, D. C., Fehrenbach, S. N., Ren, J., & Koplan, J. P. (1999). Health literacy among medicare enrollees in a managed care organization. *JAMA, 281*(6), 545–551.

German, P. S., Klein, L. E., McPhee, S. J., & Smith, C. R. (1982). Knowledge of and compliance with drug regimens in the elderly. *Journal of the American Geriatrics Society, 30,* 568–571.

Glass, R. M., Molter, J., & Hwang, M. Y. (1998). Providing a tool for physicians to educate patients: The JAMA Patient Page. *JAMA, 279*(16), 1309–1309.

Harris, J. L., & Qualls, C. D. (2000). The association of elaborative or maintenance rehearsal with age, education, reading comprehension, and verbal working memory performance. *Aphasiology, 14* (5/6), 515–526.

Harris, J. L., Rogers, W. A., & Qualls, C. D. (1998). Written language comprehension in younger and older adults. *Journal of Speech, Language, and Hearing Research, 41,* 603–617.

Hartley, J. (1999). What does it say? Text design, medical information, and older readers. In D. C. Park, R. W. Morrell, & K. Schifren (Eds.), *Processing of medical information in aging patients: Cognitive and human factors perspectives* (pp. 233–247). Mahwah, NJ: Lawrence Erlbaum Associates.

Hobbs, F. B., & Damon, B. L. (1996). Sixty-five plus (65+) in the United States. In *Current Population Reports, Special Studies* (pp. 23–190). Washington DC: U.S. Department of Health and Human Services.

Kane, R. A. (1999). Goals of home care: Therapeutic, compensatory, either, or both? *Journal of Aging and Health, 3*(11), 299–321.

Kausler, D.H. (1994). *Learning and memory in normal aging.* San Diego, CA: Academic Press, Inc.

Kemper, S. (1992). Language and aging. In F. I. M. Craik & T. A. Salthouse (Eds.), *The handbook of aging and cognition* (pp. 213–270). Hillsdale, NJ: Lawrence Erlbaum Associates.

Kempler, D. (1995). Language changes in dementia of the Alzheimer type. In R. Lubinski (Ed.), *Dementia and communication* (pp. 98–114). San Diego, CA: Singular Publishing Group.

Kempler, D., Almor, A., & MacDonald, M. C. (1998). Teasing apart the contribution of memory and language impairments in Alzheimer's disease: An online study of sentence comprehension. *American Journal of Speech-Language Pathology, 7,* 61–67.

Kintsch, W. (1980). Learning from text, levels of comprehension, or: Why anyone would read a story anyway. *Poetics, 9,* 87–98.

Lemme, B. H. (1999). *Development in adulthood,* (2nd ed). Needham Heights, MA: Allyn & Bacon.

McLaughlin, G. (1969). SMOG grading: A new readability formula. *Journal of Reading, 12* (8), 639–646.

McNeil, M. R. (1982). The nature of aphasia in adults. In N. J. Lass, L. V. McReynolds, L. Norther, & D. E. Yoder (Eds.), *Speech, language, and hearing: Volume III. Pathologies of speech and language* (pp. 692–740). Philadelphia, PA: W. B. Saunders Company.

Meyer, B. J. F., Talbot, A., Stubblefield, R. A., & Poon, L. W. (1998). Interests and strategies of young and old readers differentially interact with characteristics of texts. *Educational Gerontology, 24,* 747–771.

Morrell, R. W., & Echt, K. V. (1997). Designing written instructions for older adults to use computers. In A. D. Fisk & W. A. Rogers (Eds.), *Handbook of human factors and the older adult* (pp. 335–361). San Diego: Academic Press.

Morris, L. A., & Halpern, J. A. (1979). Effects of written drug information on patient knowledge and compliance: A literature review. *American Journal of Public Health, 69,* 47–52.

Mykityshyn, A. L. (2000). *Toward age-related training methodogies for sequence-based systems: An evaluation using a home medical device.* Unpublished master's thesis, Georgia Institute of Technology Atlanta.

National Center for Educational Statistics. (1993, September). *Adult literacy in America: A first look at the results of the National Adult Literacy Survey.* Washington, DC: Author.

National Stroke Association. (1999). [On-line]. Available: http://www.stroke.org. Accessed: 9/30/00.

Nicholas, M., Connor, L. T., Obler, L. K., & Albert, M. L. (1998). Aging, language, and language disorders. In M. T. Sarno (Ed.), *Acquired aphasia,* (3rd ed.), (pp. 413–449). San Diego: Academic Press.

Nicholas, M., Obler, L., Albert, M., & Helm-Estabrooks, N. (1985). Empty speech in Alzheimer's disease and fluent aphasia. *Journal of Speech and Hearing Research, 28,* 405–410.

Obler, L. K, Au, R., & Albert, M. L. (1995). Language and aging. In R. A. Huntley & K. S. Helfer (Eds.), *Communication in later life* (pp. 85–97). Boston: Butterworth-Heinemann.

Obler, L. K, Fein, D., Nicholas, M., & Albert, M. L. (1991). Auditory comprehension and aging: Decline in syntactic processing. *Applied Psycholinguistics, 12,* 433–52.

Park, D. C., Morrell, R. W., Frieske, C., & Kincaid, D. (1992). Medication adherence behaviors in older adults: Effects of external cognitive supports. *Psychology and Aging, 7,* 252–256.

Perlmutter, M. (1988). Cognitive potential throughout life. In J. E. Birren & V. L. Bengston (Eds.), *Emergent theories of aging* (pp. 247–267). New York: Springer.

Qualls, C. D. (1998). Figurative language comprehension in younger and older African Americans. *Dissertation Abstracts International, 59*(9–B), 4768.

Qualls, C. D. (2000). [A preliminary look at the readability of health care literature on stroke and dementia]. Unpublished data.

Qualls, C. D., Harris, J. L., & Rogers, W. A. (2000). *Understanding difficulties in written language comprehension for older adults: Memory task and genre effects.* Manuscript in preparation.

Rogers, W. A., Cabrera, E. F., Walker, N., Gilbert, D. K., & Fisk, A. D. (1996). A survey of automatic teller machine usage across the adult life span. *Human Factors, 38,* 156–166.

Rogers, W. A., Mykityshyn, A. L., Campbell, R. H., & Fisk, A. D. (2001). "Only three easy steps?" User-centered analysis of a "simple" medical device. *Ergonomics in Design, 9,* 6-14.

Rousseau, G. K., Lamson, N., & Rogers, W. A. (1998). Designing warnings to compensate for age-related changes in perceptual and cognitive abilities. *Psychology and Marketing, 15*(7), 643–662.

Salthouse, T. A. (1992). Reasoning and spatial abilities. In F. I. M. Craik & T. A. Salthouse (Eds.), *The Handbook of Aging and Cognition* (pp. 167–211). Hillsdale, NJ: Lawrence Erlbaum Associates.

Salthouse, T. (1991). *Theoretical perspectives on cognitive aging.* Hillsdale, NJ: Lawrence Erlbaum Associates.

Shadden, B. B. (1997). Language and communication changes with aging. In M. P. Cannito & D. Vogel (Series Eds.) & B. B. Shadden & M. A. Toner (Vol. Eds.), *For clinicians by clinicians: Vol. 9. Aging and communication* (pp. 135–170). Austin, TX: Pro-Ed.

Tun, P. A., & Wingfield, A. (1997). Language and communication: Fundamentals of speech communication and language processing in old age. In A. D. Fisk & W. A. Rogers (Eds.), *Handbook of human factors and the older adult* (pp. 125–149). San Diego: Academic Press.

Ulatowska, H. K., & Chapman, S. B. (1991). Neurolinguistics and aging. In D. N. Ripich (Ed.), *Handbook of geriatric communication disorders* (pp. 21–37). Austin, TX: Pro-Ed.

Vinson, B. P. (1999). *Language disorders across the lifespan.* San Diego: Singular Publishing Group.

Vogel, D., Sugar, J., & Cardillo, J. (1995, November). *Idiom explanation by older persons.* Paper presented at the annual convention of the American Speech-Language-Hearing Association, Orlando, FL.

Williams, M. V., Parker, R. M., Baker, D. W., Parikh, N. S., Pitkin, K., Coates, W. C., & Nurss, J. R. (1995). Inadequate functional health literacy among patients at two hospitals. *The Journal of the American Medical Association, 274,* 1677–1682.

Wilson, F. L. (1996). Patient education materials nurses use in community health. *Western Journal of Nursing Research, 18*(2), 195–205.

Zacks, R. T., & Hasher, L. (1994). Directed ignoring: Inhibitory regulation of working memory. In D. Dagenbach & T. H. Carr (Eds.) *Inhibitory Process in Attention, Memory, and Language* 241–264. San Diego, CA: Academic Press.

Zelinski, E. M., & Hyde, J. C. (1996). Old words, new meanings: Aging and sense creation. *Journal of Memory and Language, 35,* 689–707.

5

Medical Uncertainty and the Older Patient: An Ecological Approach to Supporting Judgment and Decision Making

Alex Kirlik
Richard Strauss
Georgia Institute of Technology

Much of the medical information available to both patient and physician alike is uncertain. A language of odds, chances, and probabilities pervades health care discussions, whether in scientific journals, in the clinic, or on the evening news. The gravity of many health care decisions only heightens awareness of these uncertainties. As a result, a patient often must cope not only with the hardship of illness, but also with doubts about when and where to turn for help, whether a current diagnosis is correct or not, or whether a prescribed course of treatment really is the best one available.

Current trends suggest that medical information will become only more uncertain in the years to come. Health care delivery and insurance industries are growing increasingly complex. Medical discoveries, procedures, and new medications appear at an accelerating rate. The Internet provides patients with ever-increasing access to huge volumes of unregulated, and sometimes even unsound, health care information and guidance (see Stronge, Walker, and Rogers, chap. 14, this volume). Paradoxically, medical advances may improve the overall quality and range of available health care options, while at the same time increasing the complexity and uncertainty of the patient's immediately local decision making environment.

Human factors interventions aimed at supporting the cognition of older adult patients must reflect the irreducible uncertainty of the health care environment in which judgments and decisions are made. Successful interventions will be guided by a knowledge of the type, frequency, and importance of the judgment and decision-making tasks older adults actually face (to ensure practical relevance), as well as by a knowledge of the cognitive resources and limitations older adults bring to these tasks (to ensure effective support). The focus in this chapter is on the use of human factors interventions to support the judgment and decision making of older patients. Complementary accounts are provided by Patel and Arocha (1999), who have discussed the effects of aging on physician cognition, and by Vroman, Cohen, and Volkman (1994), who have discussed enhancing communication and trust between physicians and older patients.

BACKGROUND AND PURPOSE

The main goal of the authors is to present and discuss a particular psychological theory and method for understanding patient decision making that they believe to be especially well suited for providing the knowledge necessary for guiding human factors interventions. Specifically, the authors discuss an ecological, or systems, approach for investigating and supporting judgment under uncertainty, motivated largely by Egon Brunswik's (1952, 1956) probabilistic functionalism and its modern extensions. An *ecological* approach, in the context of this chapter, means one that is mutually guided by expert knowledge concerning the psychology of cognition and aging, and by expert knowledge of the health care environment in which decisions and judgments actually are made. The authors' purpose is to ensure that human factors interventions are targeted to the actual judgment and decision making problems older adults face, and that these interventions are guided by psychological theory and method specifically tailored to these problems.

The authors' motivations in recommending an ecological or systems approach to human factors problems in health care are pragmatic, arising out of their experiences with design and training problems in a variety of similarly uncertain domains (Bisantz & Kirlik, 1999; Degani, Shafto, & Kirlik, 1999; Kirlik, Walker, Fisk and Nagel, 1996; Kirlik, Fisk, Walker, & Rothrock, 1998). They have learned that human factors interventions motivated mainly by domain (in this case, medical) knowledge, but by little psychological theory, are likely to be heavily opportunistic and technologically driven. Industry practitioners identify a practically relevant problem, but the choice of a technological solution often is largely opportunistic, driven by what is readily available or currently in vogue. As a result, there is little guarantee that the intervention addresses the particular cognitive factors limiting performance. It may even interfere.

On the other hand, interventions driven chiefly by psychological theory, but by little expertise in the application domain, typically fail for other reasons. Often, too little effort is taken to ensure that the psychological activities described by the theory, which may predict laboratory data well, play a major role in determining behavior in its actual, and nearly always more complex, environmental context. Charness, Kelley, Bosman, and Mottram's (1996) inability to predict the effects of word processor interface design on young and older adults' skill acquisition, using the best available theories of cognitive aging, illustrates how difficult it can be to apply theory grounded largely in laboratory data to practically relevant task environments.

Human factors researchers and practitioners are likely to be sympathetic with the authors' ecological perspective because the field has a strong tradition of focusing on the interactive human–environment system as the object of study, with the goal of creating, either through design or training, an effective match between human capabilities and environmental demands and resources. The authors also hope that health care practitioners will embrace this systems orientation for supporting the behavior of older adults because it contains resources for recognizing and representing the environmental and task knowledge that they, and not researchers on cognitive aging, often are best situated to provide. In more mature human factors application domains, such as aviation and process control, nearly all serious design and training interventions now result from problem-oriented teams. These teams contain a mix of industry professionals providing environmental and task expertise and human factors or psychology researchers and practitioners providing expertise on cognition and performance. There is no reason to expect that the problem of supporting the health care needs of an aging population will be any different. An ecological approach organizes such problem-oriented, cooperative efforts by ensuring that equal attention is paid to both the environmental and human components of the ecological system. Detailed arguments already have been made suggesting why research on cognitive aging should embrace an ecological perspective to become more practically relevant (Fisk & Kirlik, 1996), and why Brunswikian theory is particularly well suited to support the design of human factors interventions (Kirlik, in press). In this chapter we hope to flesh out what this general guidance might entail for a particular set of human factors problems in a specific application domain.

The sections that follow first motivate ecological theory and method by discussing a case study of a representative health care decision made by many older adults. The findings of that study are discussed in light of current psychological theory and previous human factors research in decision making, cognitive aging, and human error. Then follows a description of an ecological approach to supporting cognition and reducing error based on Brunswikian theory, method, and modern extensions of these ideas. Along the way, many of the more abstract concepts are illustrated through sketches of how the approach could be applied to practical issues in cognitive aging and health care. The chapter closes with the authors' recommendations for the problems that could most fruitfully be addressed by an ecological orientation to theory, method, and design.

MEDICAL INFORMATION UNCERTAINTY AND THE OLDER PATIENT

A representative and consequential decision that many older adults must make, often with the support of family and friends, is the choice of a long-term care (LTC) provider. Currently, available options for LTC include independent living (a campus-like setting with a 24-h attendant), living at home with help, assisted living, and nursing homes with varying levels of patient support.

To learn how older adults actually went about making this decision, Maloney, Finn, Bloom, and Andresen (1996) conducted in-depth interviews with 63 older adults who recently had chosen an LTC provider, along with 56 of their relatives. On the basis of these interviews, Maloney et al. (1996) found that the older adults they studied fell into one of four categories, or decision making styles: (a) scramblers, (b) reluctant consenters, (c) wake-up call decision makers, and (d) advance planners. *Scramblers* were described as those forced to respond to a crisis because of a serious injury or illness. *Reluctant consenters* were "pushed" to make a change in health care arrangements by relatives or health care professionals concerned about continued prospects for independent functioning. *Wake-up call decision makers* chose new living arrangements in response to a near crisis that could have resulted in serious injury but did not. Finally, *Advance planners* researched LTC alternatives and made plans while still healthy.

Most of the older adults participating in this study (those in the first three categories) were observed to use LTC selection strategies that were reactive, disorganized, opportunistic, and sometimes even arbitrary. Only a minority of older adults in the final, advance planners category, significantly exhibited any of the cognitive activities most researchers, such as, Yates and Patalano (1999), take to be central to decision making (e.g., identifying a set of options, evaluating these options, and so on). What factors contributed to these findings?

The answer to this question, of course, is not simple. Maloney et al. (1996) found evidence of individual differences (e.g., advance planners were described by family members as people who tended to plan extensively for most of life's major decisions, not merely LTC selection), and a tendency for older adults understandably to delay LTC decision making as long as possible. Overall, however, Maloney et al. (1996) concluded that high levels of uncertainty, from a lack of ready access to relevant information, played a major role in accounting for their data:

> This study demonstrated that for many consumers, existing LTC information is incomplete and can be hard to find. Perhaps an even greater impediment to finding appropriate LTC services, and one which is more difficult to define precisely, is the consumer's lack of familiarity with LTC issues and the choices available to address these issues. Most elderly consumers and their relatives were not familiar with the choices available to address these issues. They did not assess the needs of the elderly person and then decide which service or type of service, (i.e. home care, assisted living, adult day care, etc.), would best meet those needs. An overview of the types of options and the advantages and disadvantages of each seemed to be completely missing from the information available to respondents. (p. 10)

Clearly, not all medical decisions made by older adults are made in such a shroud of uncertainty. But the authors believe that case studies such as this should provide a wake-up call to researchers on cognitive aging who would like their research to have practical relevance. Maloney et al. (1996) began their study with the intention of gathering information on how age-related cognitive decrements might constrain decision-making effectiveness in an ecologically representative task. What they found, however, was that many of the major factors limiting the decision making of the older adults they studied were not associated with cognitive decrements at all, but instead concerned the poverty of information available in the immediately local decision environment.

Findings such as these have implications for the design of human factors interventions. Yates and Patalano (1999), for example, have provided an excellent review of the literature on judgment and decision making in older adults, emphasizing the paucity of research in this area. But the findings of Maloney et al. (1996) do not, to the authors at least, cry out for more research on aging-related cognitive decrements to support better the decisions of the older adults they studied. Instead, to the authors they suggest a need to understand better why these older adults were making such important decisions in an environment characterized by a poverty of readily available information, and a need to consider economically viable training, design, and educational remedies for improving this situation.

In other medical contexts, it might very well be the case that the effectiveness of practically relevant decisions made by older adults are mainly constrained by cognitive biases, resource limitations, and aging-related decrements in cognitive function. Some of the chapters in this volume provide such examples. But the point is that solving the problem of where to target human factors interventions requires detailed analysis of the human-environment, or ecological, system. In human factors, this type of systems or "needs and requirements" analysis (Beith, chap. 2, this volume) is necessary to ensure that interventions are targeted toward the specific aspects of the human-environment system most responsible for limiting achievement.

AN ECOLOGICAL PERSPECTIVE
ON DECISIONS AND ERRORS

The designers of human factors interventions to support the cognition of older adults in health care can benefit by heeding lessons learned in related, yet more mature, human factors applications domains. Because of the often drastic consequences of human error, research on its nature and origins has received a great deal of attention in human factors research over the past decades in domains such as nuclear power (Rasmussen, Duncan, & Leplat, 1987), commercial aviation (Wiener & Nagel, 1988), and more recently, medicine (Bogner, 1994). The following sections discuss some of the most robust and relevant findings, distinctions, and principles that have emerged from this work. This knowledge has important implications for supporting the health care–related judgment and decision making of older patients.

Error as a Systems Property

Except in situations wherein equipment failures or "acts of God" are found to play a clear role, initial investigations of industrial accidents nearly always point to the human operator (e.g., a pilot), and thus to "human error" as the ultimate cause. In most cases, however, a more complex and subtle picture begins to emerge after the passage of time allows the case to be examined in greater depth with less focus on placing blame and more focus on finding lessons to be learned. As a result of numerous such investigations, from Chernobyl to Three Mile Island to a wide variety of aviation incidents and accidents, human factors researchers have come to understand that the majority of what initially seem to be clear-cut cases of human error instead reveal long-standing organizational and design deficiencies that nurtured and fostered these errors (Rasmussen et al., 1987; Reason, 1990; Woods, Johannensen, Cook, & Sarter, 1994). Reason (1994) described many errors or failures as not active (resulting from the operator's direct interaction with the system) but rather latent:

> Unsafe acts are like mosquitoes. You can try to swat them one at a time, but there will always be others to take their place. The only effective remedy is to drain the swamps in which they breed. In the case of errors and violations, the "swamps" are equipment designs that promote operator error, bad communications, high workloads, budgetary and commercial pressures, procedures that necessitate their violation to get the job done, inadequate organization, missing barriers and safeguards. ... The list is potentially very long, but all these latent factors are, in theory, detectable and correctable before a mishap occurs. (p. xiv)

Consider you or your family's most recent brushes with the health care system. If your experiences are anything like that of the authors', the concern is not so much that any individual in the system will make an erroneous decision or negligent act, but instead that the system itself will somehow fail you. Stories of cases "slipping through the cracks" illustrate the point well, often in harrowing terms. Cognitive researchers and health care professionals interested in reducing decision-making errors must resist the tendency to see these errors as wholly resulting from deficiencies in these patients' decision making. Reconsider the findings of Maloney et al (1996) on the highly disorganized nature of many patients' activities in choosing an LTC provider. The authors' suspicion is that only a small amount of the variance in the quality of these decisions is explainable by age-related cognitive decrements. Instead, the major factors may lie in the complexity and opaque nature of the LTC environment itself, and the limited effectiveness of these patients' interactions with this environment. The low-hanging fruit for human factors interventions in decision contexts such as these can be found most readily by the efforts put forth by teams of health care and cognitive aging experts, collaborating to eliminate weak links and vulnerable points in the interactions between patients and their immediately local decision environments.

Error as Coherence Versus Correspondence Failure

In even the most supportively designed environments, some situations do inevitably crystallize into judgment or choice under irreducable uncertainty, sometimes with negative consequences. Describing any particular action (or nonaction) as an "error," however, presumes some normative yardstick against which behavior has failed to measure up. Cognitive psychologists and decision scientists have used two different types of normative criteria in the evaluation of decision making and the attribution of human error: coherence and correspondence (Hammond, 2000; Yates, 1990). Based on the philosophical distinction between coherence and correspondence theories of truth, these two criteria differ in terms of whether the focus is on measuring the match between cognitive processes and principles of rationality (coherence), or on measuring the match between the outcomes of cognition (judgments and decisions) and the actual state or demands of the external decision environment (correspondence).

These two criteria differ by whether the primary focus is placed on the adequacy of method or on the accuracy of results. Students learning long division (a type of decision procedure), for example, might be partially rewarded for using the correct algorithm (high coherence), even though they may achieve the wrong answer (low correspondence) because of simple arithmetic mistakes. Similarly, students may receive only partial credit for obtaining the correct answer (high correspondence) but failing to demonstrate correct use of the algorithm (low coherence) by neglecting to "show their work." To receive full credit, students typically must demonstrate both coherence competence and correspondence competence, although some decision situations place higher value on one type of competence than another. For example, public policy decisions may place a greater premium on rational defensibility than optimality, and thus coherence competence, whereas many personal decisions may place a greater value on getting to the right answer regardless of the method used (correspondence competence). As noted by Hammond (2000), this coherence-versus-correspondence distinction is useful for organizing findings from the history of research on judgment and decision making.

Because of its roots in economic theory, the history of the psychology of decision making has been dominated by evaluating behavior against coherence criteria. A vast amount of decision research, for example, has been motivated by subjective expected utility (SEU) theory, the notion that a person should maximize the long-term expected desirability of the consequences resulting from selecting a decision alternative (Edwards, 1954). Subjective expected utility theory provides primarily a coherence-oriented yardstick for evaluating cognition because any decision is fine, according to this theory, as long as it is internally consistent with a decision maker's preferences and a few axioms of rationality taken to be uncontroversial (e.g., that preferences should be transitive). Adequacy of method (coherence) can be evaluated against the prescriptions of SEU theory, whereas accuracy of a decision (correspondence) cannot.

Despite this limitation, and the fact that standards of rationality are hardly uncontroversial (Sen, 1993), SEU has motivated a sizable amount of behavioral research, most notably, Tversky and Kahneman's (1974) "heuristics and biases" program. Research in this tradition has been oriented traditionally toward learning how people deviate from the predictions of SEU theory and its associated prescriptions for "correct" probabilistic reasoning. Yates (1990) listed a variety of coherence failures in intuitive probabilistic reasoning, including errors in combining probabilities of conjunctive and disjunctive events, manipulating conditional probabilities, and updating beliefs according to Bayes theorem. A great deal of decision research has compared human decision making with the coherence norms provided by SEU theory, finding it to be, with few exceptions, woefully lacking. Yates and Patalano (1999) provided a review of research in this tradition, and additionally offered well-reasoned and plausible speculations about the implications of these findings for cognitive aging research.

Despite the large body of findings accumulated from research on judgment and decision making from a coherence perspective, researchers of applied theory have found it remarkably difficult to put knowledge of these cognitive biases and limitations to use in actually improving human judgment and decision making. Efforts to improve decision making through design and training interventions motivated by either prescriptive decision theory or laboratory findings on biases and heuristics have rarely proved successful (Klein, 1999). At one level, the difficulty of supporting cognition with coherence-based accounts of idealized rationality should not be hard to understand: Probability theory and SEU are major intellectual achievements, and perhaps nonspecialists should not be expected to grasp these concepts any more readily than calculus, evolution, or any other scientific theory or method. Furthermore, although SEU theory has inspired the design of many elegant laboratory experiments, it is far too cumbersome a tool for application to most of the decision problems that people face every day (Simon, 1986). Therefore, the implications of research on decision biases and related failures of coherence competence for the support of decision making in everyday contexts are unclear.

As a result, researchers of applied theory interested in creating interventions to support judgment and decision making have tended to rely more heavily on correspondence than on coherence criteria to evaluate judgment and decision making. This approach shifts the focus away from evaluating the internal coherence of cognition, and toward studying how both good and poor decisions actually are made in ecologically representative contexts. Correspondence-oriented decision research focuses on the judgment and decision strategies people use to cope with the complexity, uncertainty, and time stress characteristic of many everyday environments.

Achievement and Error as Two Sides of the Same Coin

Psychological research from this alternative, correspondence perspective has shown that many of the cognitive biases identified by comparing cognition against rational yardsticks are greatly reduced when behavior is studied in ecologically representative tasks (Cosmides & Tooby, 1996; Gigerenzer, 1991, 1998). These studies focus on the accuracy or robustness of cognition in meeting the demands of a task environment (i.e., correspondence competence).

Additionally, human factors and cognitive engineering researchers interested in supporting decision making also have come to the view that prescriptive decision theory provides few resources for supporting performers in their operational environments (Klein, Orasanu, Calderwood, & Zsambok, 1993). For example, research from the naturalistic decision-making (NDM) perspective has focused on studying how experienced professionals go about making decisions. These and related human factors studies have found little evidence that experienced performers spend much time or effort generating options, comparatively evaluating them, and so on, as prescriptive, rational, theory would demand. In contrast, these studies consistently find that a central achievement of learning is the development of "preestablished routines, heuristics, and shortcuts" (Reason, 1987, p. 468). Human factors studies in the workplace consistently find decisions emerging from seemingly effortless "recognition" (Klein, 1999), "pattern matching" (Rouse, 1983), "rule-based behavior" (Rasmussen, 1983), or by "perceptual heuristics" closely tied to concrete sources of environmental information (Kirlik, Miller, & Jagacinski, 1993; Kirlik, Walker et al., 1996).

These applied studies are guided by a correspondence perspective in which the judgment and decision making process is evaluated mainly by how adequately and robustly it meets the demands of the task environment. This evaluation is done either objectively, by performance measures that tally how well performers attain their goals, or in more subjective contexts by relying on consensual expert opinion (Klein, 1999). In this correspondence, or adaptivity-oriented research, high levels of achievement, or correspondence competence, almost always require some trade-off between decision making speed and accuracy. Heuristic strategies (i.e., "quick and dirty methods") dominated by perception, recognition, and simple rules seem to be cognition's solution to the demand for robust judgment and decision making under time stress and uncertainty.

From this correspondence-oriented, ecological perspective, achievement and error appear as two sides of the same coin. Heuristic strategies succeed, not because they are particularly intelligent or sophisticated in and of themselves, but because they cleverly and cheaply exploit information in the task environment. For example, using the old rule of thumb "red sky at night, sailors' delight; red sky in the morning, sailors' warning" displays little cognitive sophistication. Rather, the intelligence lies in how the rule supports predictions using a readily available, perceptual aspect of the environment. Moreover, when speed is an issue, this rule can be viewed to be quite intelligent, especially when contrasted with an attempt to make similar predictions using meteorological theory (requiring instead coherence competence).

However, as the simplistic nature of this rule of thumb would suggest, the efficiency of heuristic decision making comes with a price. Errors will occur when normally reliable heuristics are defeated by unusual or unnatural situations in which the assumptions underlying the simple strategies no longer hold. In atypical situations, for example, a red sky at night can be followed, not by a sailor's delight, but instead by dangerous weather. This example illustrates an ecological orientation to decision error. This perspective views error as arising largely from occasional mismatches between the human and environmental components of an interactive system that normally functions reliably.

Everything known about aging effects on cognition suggests that older adults, like experienced younger performers, tend to use judgment and decision heuristics reducing demands on attention, working memory, and self-initiated or controlled information processing (Craik & Jacoby, 1996; Yates & Patalano, 1999). To take one example, Johnson (1990) provided evidence from a task in which older adults considered fewer decision options than younger adults, but achieved comparable levels of performance. Human factors interventions for supporting the decision making of older patients are most likely to benefit by theory and method for identifying the judgment and decision heuristics older adults use to cope with uncertainty. Research also should shed light on the types of task situations that most likely will lead to error. Theory well suited to these needs is now presented.

BRUNSWIK'S PROBABILISTIC FUNCTIONALISM

Figure 5.1 depicts Brunswik's lens model, which is an integrated representation of an organism-environment system. The lens model reflects Brunswik's ecological perspective that this system, and not merely the organism, should be the appropriate object of psychological research (Brunswik, 1952). Structurally, the lens model represents the three components of an ecological system: environment, information, and organism (in the current case, a person). These components, and more specifically the relations among them, are the central theoretical constructs of Brunswik's (1956) ecological theory of probabilistic functionalism.

The Components of the Lens Model

The environmental components are depicted on the left side of Fig. 5.1, labeled as the *criterion* (Y_e), along with the *ecological validity* relations between this criterion and the *cues*, or information, available to a person. The criterion is the system's distal focus, which typically is measured as an environmental state or property (Brunswik, 1956). In a cognitive aging context, the criterion could be the risk of climbing a staircase (Wells & Evans, 1996), the exit number on a highway sign (Burns, 1998), the legitimacy of a telemarketer (Blake, 1999), or the necessity of seeking medical care (Stoller & Forster, 1994). It is crucial to note that the criterion is an environmental state, an external property *about which* judgments and decisions are made. The ecological validity relations also are environ-

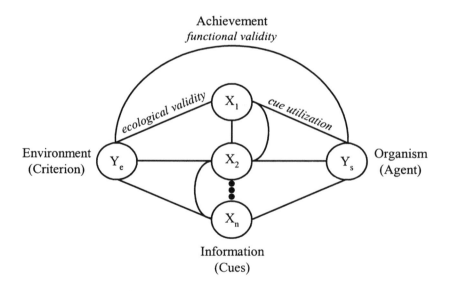

FIG. 5.1. Brunswik's lens model. Details of the model are described in the text.

mental, in that they reflect the correspondence between the environmental criterion and the cue values used for judging the criterion. In a stair-climbing context, for example, it might be the case that the actual danger or risk in climbing stairs of various designs is well predicted by cues such as riser height, tread stability, the availability of landings for resting, lighting conditions, and the like. A set of cues such as these is depicted in the figure as the collection of X_1, X_2, ... X_n values. The environmental components of the lens model are determined by expertise and analysis of the external task environment. This task is most effectively performed with the involvement of experts in the relevant application domain. In the current context, this implies that participation by health care practitioners is likely to be crucial to successful description of the environment in which older adults make medical judgments and decisions.

Reciprocally, the human components of the lens model are depicted on the right side of the Fig. 5.1 labeled as the *judgment* (Y_s), along with the *cue utilization* relations between this judgment and the available cues. Here, the judgment is the person's estimate of the value of the external criterion when viewing the available cues. To return to the stair-climbing example, judgment data would be collected by presenting people with staircases of various designs, as represented by varying cue values for riser height, tread stability, and the like, then asking them to rate the risk or danger posed by each staircase. Finally, cue utilizations are similar in spirit to their corresponding ecological validity relations, but utilizations reflect the degree to which cues are cognitively used, as well as possibly weighted and combined, in the person's judgment. In the stair-climbing case, these utilizations reflect how well cues such as riser height and tread stability are measured perceptually, and thus possibly influenced by perceptual conditions (e.g., lighting), visual acuity, and so forth.

Finally, as might be expected, the quality of judgment is measured by assessing the degree of correspondence between human judgments and criterion values. This is represented in the lens model as the arch labeled *achievement*, or *functional validity*. As can be inferred from the symmetrical structure of the model, high levels of achievement imply that the human is using the cues (the cue utilizations) in a manner that parallels the actual relations that exist between these cue values and the environmental criterion (the ecological validities).

Achievement and Error: The Lens Model Equation

Brunswik (1956) developed the lens model at a time when techniques such as regression and correlational statistics had arrived recently on the scene in psychology. The model was subsequently put on a solid mathematical foundation with the development of the lens model equation (Hursch, Hammond, & Hursch, 1964; Tucker, 1964). Reflecting the symmetry shown in Fig. 5.1, the mathematics underlying the lens model equation arises from construction of two complementary, linear regression models. The first, an environmental, regression model, describes the best linear relation between the environmental criterion and cue values. This environmental model quantifies the average correspondence, or conversely, the average degree of uncertainty, between all the cues (X_is) and the criterion (Y_e). In the staircase example, the environmental model would be found by regressing data on the actual risk, as measured either by accident data or consensual expert opinion, posed by staircases of various designs against the cue values characterizing these designs.

The second, cognitive, regression model underlying the lens model equation describes the best linear relation between human judgment and cue information. This cognitive model thus quantifies the average correspondence between all the cues (X_is) and the judgments (Y_s). For the staircase example, the cognitive model would be found by regressing data on people's judgments of the risk posed by the same staircase designs used to construct the environmental model (in other words, measures of risk perception). If the participants in such a study were well adapted to their environments, their perceptions of risk would closely correspond to the risks actually posed by various staircase designs. In such a case, it would be expected that the environmental and cognitive regression models created for this situation also would be similar. One also would expect that the achievement of these participants, measured in the lens model (the arch in Fig. 5.1) as the correlation between their judgments of risk and actual risk, to be high (near unity).

But what if achievement in some judgment or decision situation is found to be low, indicating that people are not well adapted to the demands of their environments? What, for instance, should be done if it is found that older patients' judgments about the appropriateness of various types of LTC providers, based on the information (cues) immediately available to them, are generally inaccurate? Here is where the lens model, and specifically the lens model equation, can provide specific guidance for the design of human factors interventions to improve this situation. To the question of why judgments might have a low correlation with an environmental criterion, the lens model equation, represented conceptually in Fig. 5.2 rather than mathematically, provides four possible answers.

The lens model equation decomposes achievement (or conversely, error) into four contributing factors, each of which is a relation in the ecological system. Referring to Fig. 5.2, *environmental predictability* is a measure of the goodness of fit (the multiple correlation), showing how well the environmental regression model links cue values and the criterion. High levels of environmental predictability imply highly predictable environments (e.g., staircase designs whose posed risks are well predicted by perceptually available cues) . Low levels of environmental predictability, on the other hand, suggest that the available cues do not well predict the criterion.

| Achievement = Environmental Predictability * Consistency * Knowledge + UM Behavior |

Fig. 5.2   A conceptual representation of the decomposition of the lens model equation.

Consistency is likewise a measure of the goodness of fit of the cognitive regression model. High levels of consistency imply that the behavior of the human judge is highly predictable. Low levels of consistency suggest that the available cues do not well predict the judgment that will be made (i.e., on two separate occasions a person evaluates the same staircase design, or the same LTC provider, quite differently).

The third component in Fig. 5.2 is *knowledge*. This term denotes the degree of correspondence between the environmental and cognitive models. Mathematically, knowledge is measured by correlating the predictions of the environmental and cognitive models when they are applied to the same cue sets. This correlation will be high when the human correctly models the regularities of the environment, despite any uncertainty that may be present, independently of how consistently the judge can render judgments based on this knowledge. One with expert knowledge of the LTC system, for example, is one who would know how to weight the available cues appropriately in a manner that allows optimal judgment of LTC quality, trustworthiness, and so on.

Finally, the last term in Fig. 5.2 is UM (unmodeled) *behavior*. This term arises from the possibility that a significant component of achievement may not be explained in terms of the linear regression models underlying the lens model equation. Mathematically, UM behavior is measured by correlating the residuals of the environmental and cognitive models when they are applied to the same cue sets. A high correlation indicates a shared pattern of nonlinear cue reliance in both the environment and cognition. In the vast majority of cases, UM behavior is not found to contribute significantly to achievement (Brehmer, 1994).

The mathematical form of the lens model equation is shown in Equation 1. The relation between its form and the previous conceptual presentation is detailed in Table 5.1.

$$r_a = R_e R_s G + C \sqrt{1 - R_s^2} \sqrt{1 - R_e^2}$$

Now it is possible to discuss how this modeling technique provides guidance for targeting human factors interventions to the particular aspects of the ecological system most responsible for judgment or decision-making error.

The Lens Model as a Human Factors Tool

A major use of the lens model, and its associated equation, is to identify the underlying sources of judgment errors. For example, when unmodeled behavior is negligible, Fig. 5.2 reduces to the relation shown in Fig. 5.3.

Achievement = Environmental Predictability * Knowledge * Consistency

FIG. 5.3.   A reduced form of the lens model equation.

This form of the equation emphasizes the contributions of both the human and the environment to achievement. To demonstrate the diagnostic properties of the lens model, the right side of the equation can be manipulated to show three fundamental constraints on judgment behavior: a constraint of environmental predictability (Case 1); a constraint of knowledge (Case 2), and a constraint of consistency

TABLE 5.1.

Components of the Lens Model Equation

Component	Name	Description
r_a	Achievement	Correlation between judgments and criterion
R_e	Environmental Predictability	Multiple correlation between criterion and cues
G	Knowledge	Correlation between the model of the environment and model of the human
R_s	Consistency	Multiple correlation between cues and judgments
C	UM Behavior	Correlation between the residuals of both models

(Case 3). Each case is discussed in turn. Note that as correlations, environmental predictability, knowledge, and consistency can never be greater than unity.

In Case 1, when both knowledge and consistency are perfect, and thus each equal to unity, Fig. 5.3 reduces to the relation shown in Fig. 5.4.

| Achievement = Environmental Predictability |

FIG. 5.4.   Case 1.

Figure 5.4 depicts environmental predictability as a constraint to achievement, even when the human is perfectly knowledgeable about the judgment task and demonstrates perfect consistency in judgment behavior. Consider again an older adult confronted with the task of judging the risk of climbing the neighbor's patio staircase. To make this judgment, the adult relies on three cues: the number of stairs, the height of the risers, and the stability of the treads. As assumed in Case 1, the adult is perfectly knowledgeable in making such judgments, and demonstrates perfectly consistent judgment behavior (e.g., staircases with fractured treads are always judged as high risk). Yet this adult may still make judgment errors because the available information may be uncertain (e.g., the unsteadiness of a tread often is hidden). In such cases, human factors efforts are needed to improve the predictability of the task environment by increasing the availability of informative judgment or decision cues (Stewart, 1999).

In Case 2, when both environmental predictability and consistency are perfect, Fig. 5.3 reduces to the relation shown in Fig. 5.5.

Figure 5.5 depicts knowledge as a constraint to achievement, even when the environment is perfectly predictable and the human demonstrates perfectly consistent judgment behavior. Consider a patient on a regimen of multiple prescriptions who must safely combine these medications throughout the day. Consider, however, that unlike the situation in Case 1, the environment is perfectly predictable given the available information (the correct combinations that avoid unsafe drug interactions are well established), and the patient makes perfectly consistent judgments (always combining the blues with the reds). Yet under even these conditions, the patient's strategy might be consistently wrong. Here, interventions should be directed toward improving the patient's knowledge of the environmental constraints defining good judgment.

| Achievement = Knowledge |

FIG. 5.5.   Case 2.

Finally, in Case 3, when both environmental predictability and knowledge are perfect, Fig. 5.3 reduces to the relationship shown in Fig. 5.6.

Figure 5.6 depicts consistency as a constraint to achievement, even when the environment is perfectly predictable and the human is perfectly knowledgeable about the judgment task. Using the medication example, Case 3 finds the patient in a perfectly predictable environment with perfect knowledge of the judgment task, but with an inability to consistently transform this knowledge into correct judgments (e.g., one drug has the side effect of combining with particular food to produce drowsiness). As in Case 2, Case 3 directs human factors interventions toward the human, but with the intent of improving consistency of execution. Interventions in the form of automated reminders or augmented feedback (Cooksey, 1996) could be warranted.

| Achievement = Consistency |

FIG. 5.6.   Case 3.

Finally, it should be emphasized that each of the preceding cases above is an idealization illustrating how the lens model allows performance or error to be decomposed into its contributing sources. In any actual application, it is likely that some combination of less than perfect knowledge, consistency, and environmental predictability will contribute to error. The lens model allows the relative contributions of these factors to be measured, however, and thus assists in targeting interventions toward the major causes of inadequate human performance.

Applications of the Lens Model to Cognitive Aging

The lens model has been used to explore a diverse set of problems including clinical judgment, conflict resolution, interpersonal learning, and expertise (for a partial review see Brehmer & Brehmer, 1988). In one of the few studies using the lens model to investigate age-related differences, Chasseigne, Grau, Mullet, and Cama (1999; see also Chasseigne, Mullet, & Stewart, 1997) challenged the typical finding that older adults perform worse than younger adults in a learning task (Salthouse, 1994a). In their task, 220 participants from four age groups, encompassing an age range from 18 to 90 years, were asked to learn the cue–criterion relations underlying a weather forecasting system. The task was rather straightforward in that all four cues had positive correlations with the criterion, and the cue intercorrelations (a measure of information redundancy) all were zero. In contrast to the typical deterministic tasks used to investigate cognitive aging, these researchers used the leverage of the lens model to introduce five levels of environmental predictability (see Table 5.1). Their data and lens model analysis found no significant uncertainty x age interactions, and nearly equivalent levels of achievement between older and younger participants across all five levels of uncertainty, a finding that might come as a surprise to those acquainted with the cognitive aging literature.

How should the lack of an age decrement be interpreted in that study given the range of well-documented age decrements described in the cognitive literature (Chasseigne et al., 1999; Salthouse, 1994b)? The answer, of course, must lie in the nature of the demands various tasks make on various aspects of cognitive functioning. As discussed by Hammond (2000), instruments such as the Raven Progressive Matrices test, which typically show substantial age-related decrements in fluid intelligence, weigh much more heavily on what the current authors have called coherence competence than on correspondence competence because they measure "analytical intelligence" and the ability to "reason" (Carpenter, Just & Shell, 1990, p. 404). The results obtained by Chasseigne et al. (1999) suggest that care should be taken in assuming that age-related cognitive decrements identified by such tests reveal much about how well older adults can perform primarily correspondence-oriented tasks, such as making accurate judgments in environments described by straightforward cue–criterion relations (cues correlated positively with the criterion and lacking cue intercorrelations). To prevent premature or overgeneralization of findings, the task conditions under which the findings were obtained must be specified precisely, ideally through modeling.

One of the strengths of the lens model, and ecological approaches generally, is in the resources provided for environmental modeling to aid in meeting this goal. For instance, research conducted in the lens model tradition has discovered that people have a harder time learning to perform judgment tasks in environments wherein cues are not all correlated positively with the environmental criterion (Brehmer & Brehmer, 1988). Einhorn, Kleinmuntz, and Kleinmuntz (1979) noted that the vast majority of cue–criterion relations characterizing the physical environment (area, volume, distance traveled), are positive and monotonic in their arguments and thus, whether through evolution or development, people may have a tendency to assume positive cue–criterion relations unless they discover them to be otherwise. Because a variety of studies have demonstrated that older adults have more difficulty than younger adults in altering a previously developed schema (Craik & Jacoby, 1996; Rice & Okun, 1994) or habit (Luchins & Luchins, 1994), it might be expected that older adults would find judgment tasks requiring the discovery and use of cues negatively correlated with an criterion especially troublesome. An additional experiment by Chasseigne et al. (1997) suggested this indeed to be the case, that older adults had more difficulty learning a negatively correlated task than younger adults.

These studies on cognitive aging, conducted within a Brunswikian orientation, have highlighted the sensitivity of performance to variations in task conditions, demonstrating how these conditions can be described precisely given a model with resources for representing both the human and environmental components of an ecological system. To the current authors at least, such an approach seems essential for identifying the task conditions that most likely will lead to judgment and decision errors in older adults, and for ensuring reliable generalization of findings across experimental task conditions or from laboratory situations to practically relevant contexts.

ECOLOGICAL RATIONALITY

The lens model portrays judgment and decision making in simple, heuristic terms: Find some cues that seem to work; weight and combine them; and render a response. Although this discussion has been presented as if these cues are always perceptual in nature, this need not always be the case. Often judgments are based on cues drawn from memory, and because of retrievability and working memory limitations, it can be expected that such judgments might be based on fewer cues than would be the case if the task environment were perceptually available.

In their recent book, Gigerenzer and the ABC Research Group (1999) introduced and explored the possibility that judgment and decision heuristics even simpler than those assumed by the lens model might underlie a great majority of memory-based judgments. These researchers noted that the cognitive demands of compensatory cue weighting and integration may be unreasonable under conditions requiring quick judgments with limited knowledge. As a remedy, they propose a set of even simpler, noncompensatory heuristics that are both fast (require little computation) and frugal (use only a few cues). Their claim is that such heuristics demonstrate a higher degree of ecological rationality than the compensatory strategies assumed by the lens model. That is, they suggest that their fast and frugal heuristics provide an even more robust match between the demands imposed by the environment and human cognition. One of these heuristics is discussed here, because it is an intriguing possibility that as age-related cognitive decrements appear, older adults may tend to use simpler, heuristic strategies making reduced information processing demands.

The Recognition Heuristic

As an example of what they call "ignorance-based decision making," (i.e. where knowing less can be an asset), Goldstein and Gigerenzer (1999) introduced the recognition heuristic. The rules underlying the heuristic are straightforward: Search for the first cue that is recognized for one object but not for the other; stop searching once the cue is found; and then select the object with the recognized cue as the object with the higher value. Although this sounds like a vastly oversimplified judgment heuristic, its simplicity is deceptive. Borges, Goldstein, Ortmann, and Gigerenzer (1999) conducted a series of studies to test the power of the recognition heuristic for selecting stocks. They asked both laypersons (pedestrians in two large cities) and experts (finance and economics graduate students) to select companies they recognized from a list of 798 companies. They found that stock portfolios based on the recognized companies outperformed portfolios based on the unrecognized companies, with one measure showing the returns from recognized stocks to be three times higher on the average than returns from unrecognized stocks. Moreover, portfolios based on the 10 most selected companies by a set of laypersons, which had an average recognition rate of .80, significantly outperformed an equivalent portfolio selected by a set of experts, which had an average recognition rate of only .48. Indeed, a judgment strategy based on the recognition heuristic can deliver some provocative results.

Fast and Frugal Heuristics and the Older Adult

Given the recency of studies on fast and frugal heuristics by Gigerenzer et al. (1999), no applications of these ideas to age-related issues yet exist. Converging evidence from existing research suggests, however, that a search for simple, memory-based judgment heuristics possibly used by older adults may be a fruitful area of study. For example, several lines of research have indicated that older adults have reduced working memory capacity (Salthouse, 1990). This finding would seem, to the current authors at least, to increase the ecological rationality of simple judgment and decision heuristics. Therefore, one immediate question is whether there is in fact an age-related difference in reliance on simple heuristics based on very small numbers of cues. It is a likely hypothesis that such reliance would increase with age. Consideration should be given as well to Yates and Patalano's (1999) suggestion that the cognition of older adults can be characterized more by reactive and automatic decision making, and less by memory-intensive, effortful analysis. Extensions of Brunswik's probabilistic functionalism to the realm of memory-based judgments may provide resources for identifying how older adults might be able to maintain acceptable performance despite these limitations, and also may highlight those judgments and decisions that older adults probably should not make solely on the basis of cues retrieved from memory, but instead with the aid of well-designed environmental support.

CONCLUSION

Older patients often make health care judgments and decisions in highly uncertain environments. The authors presented an ecological approach to the investigation of patient judgment, decision making, and error that they believe holds promise for guiding the design of human factors interventions. The theory and method is based on Brunswik's (1956) probabilistic functionalism and modern extensions of his systems-oriented, ecological psychology. This approach focuses on analyzing the environments in which judgments and decisions are made, the cognitive resources and strategies people bring to these tasks, and the degree of match or mismatch between these environmental and human elements of the ecological system.

Most successful design and training interventions in more mature human factors domains now result primarily from teams of industry professionals and psychological researchers bringing together expertise on the environmental and human components of the ecological system. An ecological approach recognizes and embraces both types of expertise as necessary for performing practically relevant research. The primary reason why an ecological approach is particularly well suited to guiding human factors interventions is because it requires that both cognition and environment be represented by abstractions (models) at similar levels of detail and sophistication.

Brunswik's (1956) lens model provides an excellent illustration of the symmetry underlying an ecological approach, but it should be pointed out that the principle of taking the integrated, human-environment system as the unit of psychological analysis transcends any particular modeling technique. Those involved with human factors problems other than those supporting judgment and decision making will require representational techniques other than the lens model to describe the environmental components of their ecological systems of interest (e.g., for modeling the display/control structure of medical devices, telemedicine interfaces, and so on). Studies based on models representing both internal cognition and the external task environment enable practitioners to target human factors interventions toward those features of the world most likely to limit performance, and researchers to justify generalization of theories or findings across conditions using a logic of representativeness rather than informal argument or trial and error.

REFERENCES

Bisantz, A. M., & Kirlik, A. (1999). Adaptivity and rule verification tasks: An empirical investigation. *International Journal of Cognitive Ergonomics, 4*(1) 1–18.

Blake, K. (1999, August). More elderly sweepstakes abuse. *Consumer's Research Magazine, 82*, 38.

Bogner, M. S. (1994). *Human error in medicine*. Hillsdale, NJ: Lawrence Erlbaum Associates.

Borges, B., Goldstein, D. G., Ortmann, A., & Gigerenzer, G. (1999). Can ignorance beat the stock market? In G. Gigerenzer, P. M. Todd, & the ABC Research Group (Eds.), *Simple heuristics that make us smart* (pp. 59–74). Oxford, England: Oxford University Press.

Brehmer, A., & Brehmer, B. (1988). What have we learned about human judgment from thirty years of policy capturing? In B. Brehmer & C. R. B. Joyce (Eds.), *Human judgment: The SJT view* (pp. 75–114). Amsterdam: North-Holland.

Brehmer, B. (1994). The psychology of linear judgment models. *Acta Psychologica, 87*, 137–154.

Brunswik, E. (1952). The conceptual framework of psychology. In the *International Encyclopedia of Unified Science* (Vol. 1, pp. 4–102). Chicago: University of Chicago Press.

Brunswik, E. (1956). *Perception and the representative design of psychological experiments*. Berkeley, CA: University of California Press.

Burns, P. C. (1998). Wayfinding errors while driving. *Journal of Environmental Psychology, 18*(2), 209–217.

Carpenter, P. A., Just, M. A., & Shell, P. (1990). What one intelligence test measures: A theoretical account of processing in the Rave Progressive Matrices Test. *Psychological Review, 92,* 404–431.

Charness, N., Kelley, C., Bosman, E., & Mottram, M. (1996). Cognitive theory and word processing training: When prediction fails. In W. A. Rogers, A. D. Fisk, & N. Walker (Eds.), *Aging and skilled performance: Advances in theory and applications* (pp. 221–240). Mahwah, NJ: Lawrence Erlbaum Associates.

Chasseigne, G., Grau, S., Mullet, E., & Cama, V. (1999). How well do elderly people cope with uncertainty in a learning task? *Acta Psychologica, 103,* 229–238.

Chasseigne, G., Mullet, E., & Stewart, T. R. (1997). Aging and multiple cue probability learning: The case of inverse relationships. *Acta Psychologica, 97*(3), 235–252.

Cooksey, R. W. (1996). *Judgment analysis: Theory, methods, and applications.* San Diego, CA: Academic Press.

Cosmides, L., & Tooby, J. (1996). Are humans good intuitive statisticians after all? Rethinking some conclusions from the literature on judgment under uncertainty. *Cognition, 58,* 1–73.

Craik, F. I. M., & Jacoby, L. L. (1996). Aging and memory: Implications for skilled performance. In W. A. Rogers, A. D. Fisk, & N. Walker (Eds.), *Aging and skilled performance: Advances in theory and applications* (pp. 113–137). Mahwah, NJ: Lawrence Erlbaum Associates.

Degani, A., Shafto, M., & Kirlik, A. (1999). Modes in human-machine systems: Review, classification, and application." *International Journal of Aviation Psychology,* (2), 125–138.

Edwards, W. (1954). The theory of decision making. *Psychological Review, 51,* 380–417.

Einhorn, H. J., Kleinmuntz, D. N., & Kleinmuntz, B. (1979). Linear regression *and* process-tracing models of judgment. *Psychological Review, 86,* 465–485.

Fisk, A. D., & Kirlik, A. (1996). Practical relevance and age-related research: Can theory advance without application? In W. A. Rogers, A. D. Fisk, & N. Walker (Eds.), *Aging and skilled performance: Advances in theory and applications* (pp. 1–15). Mahwah, NJ: Lawrence Erlbaum Associates.

Gigerenzer, G. (1991). How to make cognitive illusions disappear: Beyond heuristics and biases. In W. Stroebe & M. Hewstone (Eds.), *European review of social psychology,* (Vol. 2., pp. 83–115). New York: John Wiley & Sons.

Gigerenzer, G. (1996). Rationality: Why social context matters. In P. B. Baltes & U. M. Staudinger (Eds.), *Interactive minds: Life-span perspectives on the social foundation of cognition* (pp. 319–346). Cambridge, MA: Cambridge University Press.

Gigerenzer, G. (1998). Ecological intelligence: An adaptation for frequencies. In D. D. Cummins & C. Allen (Eds.), *The evolution of mind* (pp. 9–29). New York: Oxford University Press.

Gigerenzer, G., Todd, P. M., & the ABC Research Group (1999). *Simple heuristics that make us smart.* Oxford, England: Oxford University Press.

Goldstein, D. G., & Gigerenzer, G. (1999). The recognition heuristic: How ignorance makes us smart. In G. Gigerenzer, P. M. Todd, & the ABC Research Group (Eds.), *Simple heuristics that make us smart* (pp. 37–58). Oxford, England: Oxford University Press.

Hammond, K. R. (2000). *Judgments under stress.* New York: Oxford University Press.

Hursch, C. J., Hammond, K. R., & Hursch, J. L. (1964). Some methodological considerations in multiple-cue probability studies. *Psychological Bulletin, 71,* 42–60.

Johnson, M. M. (1990). Age differences in decision making: A process methodology for examining strategic information processing. *Journal of Gerontology, 45,* 75–78.

Kirlik, A. (in press). Human factors. In K. R. Hammond & T. Stewart (Eds.), *The essential Brunswik: Beginnings, explications, and applications.* Cambridge: Oxford University Press.

Kirlik, A., Fisk, A. D., Walker, N., & Rothrock, L. (1998). Feedback augmentation and part-task practice in training dynamic decision-making skills. In J. A. Cannon-Bowers & E. Salas (Eds.), *Making decisions under stress: Implications for individual and team training.* (pp. 91–113). Washington, DC: American Psychological Association.

Kirlik, A., Miller, R. A., & Jagacinski, R. J. (1993). Supervisory control in a dynamic uncertain environment: A process model of skilled human-environment interaction. *IEEE Transactions on Systems, Man, and Cybernetics, 23*(4), 929–952.

Kirlik, A., Walker, N., Fisk, A. D., & Nagel, K. (1996). Supporting perception in the service of dynamic decision making. *Human Factors, 38*(2), 288–299.

Klein, G. A. (1999). Applied decision making. In P. A. Hancock, (Ed.). *Handbook of perception and cognition: Human performance and ergonomics* (pp. 87–108). New York: Academic Press.

Klein, G. A., Orasanu, J., Calderwood, R., & Zsambok, C. E. (Eds.). (1993). *Decision making in action: Models and methods.* Norwood, NJ: Ablex Corporation.

Luchins, A. S., & Luchins, E. H. (1994). The water jar experiments and Einstellung effects: II. Gestalt psychology and past experience. *Gestalt Theory, 16*(4), 205–259.

Maloney, S. K., Finn, J., Bloom, D. L., & Andresen, J. (1996). Personal decision-making styles and long-term care choices. *Healthcare Financing Review, 18*(1), 141–155.

Patel, V., & Arocha, J. F. (1999). Medical expertise and cognitive aging. In D. C. Park, R. W. Morrell, & K. Shifren (Eds.), *Processing of medical information in aging patients: Cognitive and human factors perspectives* (pp. 127–143). Mahwah, NJ: Lawrence Erlbaum Associates.

Rasmussen, J. R. (1983). Skills, rules and knowledge: Signals, signs, symbols, and other distinctions in human performance models. *IEEE Transactions on Systems, Man, and Cybernetics, SMC-13*(3), 257–266.

Rasmussen, J. R., Duncan, K., and Leplat, J. (Eds.). (1987). *New technology and human error.* New York: Wiley.

Reason, J. T. (1987). Cognitive aids in process environments: Prostheses or tools? *International Journal of Man-Machine Studies, 27,* 463–470.

Reason, J. T. (1990). *Human error.* New York: Cambridge University Press.

Reason, J. T. (1994). Foreword. In M. S. Bogner (Ed.), *Human error in medicine* (pp. vii– xv).

Rice, G. E., & Okun, M. A. (1994). Older readers' processing of medical information that contradicts their beliefs. *Journal of Gerontology, 49*(3), 119–128.

Rouse, W. B. (1983). Models of human problem solving: Detection, diagnosis, and compensation for system failures. *Automatica, 19,* 613–625.

Salthouse, T. A. (1990). Working memory as a processing resource in cognitive aging. *Developmental Review, 10,* 101–124.

Salthouse, T. A. (1994a). Aging associations: Influence of speed on adult age differences in associative learning. *Journal of Experimental Psychology: Learning, Memory, and Cognition, 20*(6), 1486–1503.

Salthouse, T. A. (1994b). The nature of the influence of speed on adult age differences in cognition. *Developmental Psychology, 30,* 240–259.

Sen, A. (1993). Internal consistency of choice. *Econometrica, 61*(3), 495–521.

Simon, H. A. (1986). Alternative visions of rationality. In H. R. Arkes & K. R. Hammond (Eds.), *Judgment and decision making: An interdisciplinary reader* (pp. 97–113). Cambridge, England: Cambridge University Press.

Stewart, T. R. (1999). Improving reliability of judgmental forecasts. In J. S. Armstrong (Ed.), *Principles of forecasting: A handbook for researchers and practitioners.* Norwell, MA: Kluwer.

Stoller, E. P., & Forster, L. E. (1994). The impact of symptom interpretation on physician utilization. *Journal of Aging and Health, 6*(4), 507–534.

Tucker, L. R. (1964). A suggested alternative formulation in the developments of Hursch, Hammond, and Hursch, and by Hursch, Hammond, and Todd. *Psychological Review, 71*(6), 528–530.

Tversky, A., & Kahneman, D. (1974). Judgment under uncertainty: Heuristics and biases. *Science, 185*, 1124–1131.

Vroman, G., Cohen, I., & Volkman, N. (1994). Misinterpreting cognitive decline in the elderly: Blaming the patient. In M. S. Bogner (Ed.), *Human error in medicine* (pp. 93–122). Hillsdale, NJ: Lawrence Erlbaum Associates.

Wells, N. M., & Evans, G. W. (1996). Home injuries of people over age 65: Risk perceptions of the elderly and of those who design for them. *Journal of Environmental Psychology, 16*(3), 247–257.

Wiener, E., & Nagel, D. (Eds.). (1988). *Human factors in aviation.* San Diego: Academic Press.

Woods, D. D., Johannesen, L., Cook, R. I., & Sarter, N. (1994). *Behind human error: Cognitive system, computers, and hindsight.* Wright-Patterson AFT, OH: Crew Systems Ergonomics Information Analysis Center.

Yates, J. F. (1990). *Judgment and decision making.* Englewood Cliffs, NJ: Prentice-Hall.

Yates, J. F., & Patalano, A. L. (1999). Decision making and aging. In D. C. Park, R. W. Morrell, & K. Shifren (Eds.), *Processing of medical information in aging patients: Cognitive and human factors perspectives* (pp. 31–54). Mahwah, NJ: Lawrence Erlbaum Associates.

6

Exercise, Aging, and Cognition: Healthy Body, Healthy Mind?

Arthur F. Kramer
Sowon Hahn
Edward McAuley
Neal J. Cohen
Marie T. Banich
Cate Harrison
Julie Chason
Richard A. Boileau
Lynn Bardell
Angela Colcombe
Beckman Institute, University of Illinois at Urbana-Champaign,
Eli Vakil
Bar-Ilan University, Israel

The main goal of this chapter is to examine the relation between improvements in aerobic fitness and the cognitive function of sedentary older adults. An attempt has been made to accomplish this goal in two ways: by a critical review of the literature and by presenting the results of a recent study conducted in the authors' laboratory. The main hypothesis for this study, motivated by a critical analysis of the literature on fitness, aging, and cognition, was that improvements in aerobic fitness would result in selective improvements in cognitive function. More specifically, the authors predicted that improvements would be observed in executive control processes (i.e., planning, scheduling, coordination, inhibition), that is, those processes subserved largely by the frontal and prefrontal regions of the brain.

Before beginning a review of the literature on the relation among aging, fitness, and cognition, it might be helpful to describe explicitly the logic that underlies research in this domain. First, it is generally the case that cerebrovascular and cardiorespiratory insufficiency increases between young adulthood and old age (Marchal, 1992; Meyer, Kawamura, & Terayama, 1994). Second, declines in cognitive function can be caused by decreases in cardiorespiratory and cerebrovascular efficiency (Emery, Schein, Hauck, & McIntyre, 1998; Dustman, Emmerson, & Shearer, 1994). Third, cardiovascular conditioning has been demonstrated to enhance cerebrovascular sufficiency by increasing aerobic capacity or cardiac output through increased stroke volume and oxygen extraction in older humans (Boutcher & Landers, 1998) to promote the development of new capillary networks in the brains of old rats (Jones, Hawrylak, Klintsova, & Greenough, 1998), to enhance cortical high-affinity choline uptake and increase dopamine receptor density in the brains of old rats (Fordyce & Farrar, 1991), and to increase brain derived neurotrophin factor gene expression in rats (Neeper, Gomez-Pinilla, Choi, & Cotman, 1995). Finally, using this chain of logic and supporting data as a basis, it has been speculated that cardiorespiratory conditioning will enhance cerbrovascular efficiency, which in turn will serve to reduce age-related decline in cognitive function (Blumenthal & Madden, 1988; Cotman & Neeper, 1996; Dustman et al., 1994; Hawkins, Kramer, & Capaldi, 1992).

AGING, FITNESS, AND COGNITION

Early cross-sectional studies on the potential benefits of aerobic fitness for cognitive processing in older adults focused on simple and choice reaction-time (RT) tasks and provided evidence that fitness may be associated with slower age-related declines. In a classic study, Spirduso (1975) found that older racquet sportsmen were significantly faster on simple, choice and movement response times than older nonexercisers. In a follow-up study, Spirduso and Clifford (1978) found similar results for runners compared with nonrunners (see also Offenbach, Chodzko-Zajko, & Ringel, 1990; Stones & Kozma, 1988). Additionally, exercisers have been found to outperform nonexercisers on tasks such as reasoning, working memory, Stroop, Trails-B, symbol digits, vigilance monitoring, and fluid intelligence tests (Abourezk & Toole, 1995; Bunce, Barrowdough, & Morris, 1996; Cook et al., 1995; Dustman, Emerson, & Shearer 1994). However, differences in performance on seemingly similar tasks between lifetime exercisers and non-exercisers have not always been found. Reports of failure to find beneficial effects of exercise on the performance of simple and choice reaction-time tasks, short-term memory and digit span tasks, and measures of somatosensory thresholds have been reported (Abourezk & Toole, 1995; Clarkson-Smith & Hartley, 1990; Dustman et al., 1990; van Boxtel et al., 1997).

Although the preceding research has, in general, found beneficial effects of exercise on the cognitive processes of older adults, the cross-sectional nature of these studies complicates their interpretation. Therefore, the positive effects may reflect a genetic predisposition of the exercisers toward fast and accurate responding rather than a benefit of aerobic fitness achieved through exercise. A number of researchers have circumvented the problem of self-selection at least partially by using longitudinal exercise interventions.

Dustman et al. (1984) found improvements in the performance of a number of tasks including critical flicker fusion, digit symbol, and Stroop after a 4-month exercise program. These improvements were specific to an aerobic exercise group exhibiting a considerable improvement in cardiovascular function (i.e., 27% improvement in maximum oxygen consumption [VO_2max]). Subjects who had participated in a strength and flexibility program and those in a nonexercise control group did not show improvements in performance across test administrations. However, consistent with previous longitudinal studies, there were also a variety of tasks in which aerobic exercise failed to have a beneficial effect. These tasks included choice RT and culture fair IQ tasks.

Rikli and Edwards (1991) found that a 3-year aerobic fitness program served to eliminate declines in choice reaction-time performance that were observed for a nonexercise control group. Interestingly, fitness-related performance sparing was not observed for a simple RT task.

Finally, Hawkins et al. (1992) found significant improvements in the dual-task performance of elderly adults who participated in a 10-week aerobic fitness program while failing to observe any improvement in the dual-task conditions for a nonexercise control group. Both the exercise and nonexercise control groups showed comparable performance improvements on the single tasks, presumably because of practice. Taken together, the results of these longitudinal studies (see also Chodzko-Zajko & Moore, 1994; Moul, Goldman, & Warren, 1995) are supportive of selective improvements in a number of cognitive processes with short-term programs of aerobic exercise.

Other longitudinal studies have failed to find aerobic exercise benefits in human information processing. Blumenthal and Madden (1988) reported that both aerobic and anaerobic exercise groups improved their performance on a memory search task across test administrations that spanned 12 weeks. Failure to find a performance improvement attributable to the aerobic exercise program may be the result of several factors. First, the subjects were relatively young, ranging in age from 30 to 58 years. Previous studies (Barry et al., 1966; Dustman et al., 1984; Elsayed et al., 1980) have found more robust effects of exercise on the cognitive processes of older adults. Second, the participants were relatively fit before the exercise program with an average VO_2max of 34 ml/kg/min. Third, improvement in cardiovascular function was markedly less than the improvement reported by Dustman et al. (1984). Finally, rehearsal time for the memory task was uncontrolled. Subjects were permitted to view the memory set as long as they wished. Therefore, it is conceivable that aerobic exercise may have had a beneficial effect on encoding processes.

Madden, Blumenthal, Allen, and Emery (1989) reexamined the effects of aerobic exercise on memory search performance. Older (ages 60 to 83 years) and less fit subjects performed a memory search both separately and in conjunction with a secondary auditory task. The secondary task was included in an effort to determine whether more attentionally demanding tasks would benefit from short term programs of exercise. There was no difference between aerobic exercise group and control groups in memory performance. The cardiovascular improvement exhibited by the aerobic group again was smaller than that reported by Dustman et al. (1984) (see also Blumenthal et al., 1991; Hill, Storandt, & Malley, 1993; Panton, Graves, Pollock, Hagberg, & Chen, 1990).

In summary, the literature suggests that although a lifetime of aerobic exercise may help to preserve a selective subset of cognitive capabilities in older adults, the benefits of exercise intervention programs are much more equivocal. A plausible interpretation of the literature concerns the nature of the cognitive processes that have been examined as well as the exercise interventions. A comparison of the fitness levels reported for the postexercise groups in the longitudinal studies and those reported for the lifetime exercisers clearly suggests that high levels of fitness are achieved after years rather than months of training. Therefore, it might be unreasonable to expect that brief programs of exercise will have beneficial effects on a wide variety of cognitive processes. Instead, it is conceivable that short-term exercise benefits might be restricted to those processes that have the most room for improvement, that is these processes (i.e., executive control processes) most susceptible to aging. The next section describes the literature on age-related differences in executive control processes and the brain structures that support them.

AGING AND EXECUTIVE CONTROL

In the 1990s, there was a renewed interest among cognitive psychologists and cognitive aging researchers in executive control functions. Such functions are concerned with the selection, inhibition, scheduling, and coordination of the computational processes responsible for perception, memory, and action. The interest in the executive control of cognitive processes has been reflected in the development of cognition models that reserve an important role for executive control functions (Baddeley, 1992; Shallice, 1994), and in the detailed empirical examination of a subset of executive control processes of young and old adults (Allport, Styles, & Hsieh, 1994; Kramer et al., 1999b, 2000; Rogers & Monsell, 1995; Verhaeghen, Kliegl, & Mayr, 1997).

In his recent critical review of the literature on the neuroanatomy, neurophysiology, and neuropsychology of aging, West (1996) concluded that relatively strong evidence exists for the frontal lobe hypothesis of cognitive aging. This hypothesis suggests that older adults are disproportionately disadvantaged in tasks that rely heavily on cognitive processes (i.e., executive control processes) supported by the frontal and prefrontal lobes of the brain. Indeed, there is a great amount of evidence to suggest that morphologic and functional changes in brain activity do not occur uniformly during the process of normal aging. Researchers have reported substantially larger reductions of gray matter volume in association areas of cortex, particularly in the prefrontal and frontal regions, than in sensory cortical regions (Raz, in press). Studies of functional brain activity using positron emission tomography (PET) have reported similar trends, with prefrontal regions showing substantially larger decreases in metabolic activity than sensory areas of cortex (Azari et al., 1992; Salmon et al., 1992).

These data on the structure and function of the aging brain are consistent with numerous reports of large and robust age-related deficits in the performance of tasks supported largely by the frontal and prefrontal regions of the cortex, as compared with the relatively small age-related deficits in nonfrontal lobe tasks (Daigneault, Braun, & Whitaker, 1992; Shimamura & Jurica, 1994). Indeed, many of the tasks subserved, in large part, by the frontal lobes involve processes associated with executive control functions such as the selection, control, and coordination of computational processes responsible for perception and action. For example, large age-related deficits generally have been reported in adults required to perform two or more tasks at the same time or to shift emphasis rapidly among tasks (Kramer et al., 1999c; Rogers et al., 1994). Functional magnetic resonance imaging (fMRI) and PET studies have shown enhanced activation of regions of the prefrontal and frontal cortices when two tasks are performed together, but not when they are performed separately (D'Esposito et al., 1995). Verhaeghen et al. (1997) also reported that reliably larger age-related performance decrements are observed for tasks that require coordinative operations (e.g., mental arithmetic operations in which a product must be held in working memory as other computations are performed) than for tasks that require sequential operations (i.e. mental arithmetic operations that do not require storing and retrieving products from working memory while carrying out arithmetic operations).

Older adults also show disproportionate age-related performance deficits in tasks the require the inhibition of prepotent responses such as the antisaccade paradigm (Kramer et al., 2000) and the Stroop task (Brink & McDowd, 1999; Hartley, 1993). The antisaccade paradigm requires that subjects quickly shift their eyes in the direction opposite that of a stimulus. That is, they are to inhibit a reflexive eye movement to the stimulus and move their eyes in the opposite direction. Performance of this task is extremely difficult for patients with lesions in the dorsolateral prefrontal cortex and frontal eye fields, and has been shown to activate these brain regions in PET studies (Corbetta, 1998). The Stroop task requires subjects to name the color of the ink used for a word presented. Response times and errors are elevated when the word names an incongruent color, particularly for older adults.

In summary, the abbreviated review of the cognition and aging literature suggests relatively strong support for the hypothesis that older adults are more disadvantaged by tasks that rely heavily on executive control processes (and the frontal and prefrontal cortices) than by tasks that rely on component task processes (i.e., perceptual and action-related processes). Although it is clear that older adults often are slower than young adults in performing many tasks (Salthouse, 1996), tasks that rely heavily on the executive control processes of planning, scheduling, coordinating, and inhibition of prepotent responses appear to suffer a disproportionate cost in terms of performance as well as the brain structures and functions that support them.

AGING, FITNESS, AND COGNITION REVISITED

The reviewed literature on cognition and aging suggests that older adults are disproportionately disadvantaged when they must perform tasks that rely heavily on executive control processes and the brain regions supporting them. In this section, the literature on aging, fitness, and cognitive function is reexamined in light of the cognition and aging literature. The main question addressed here is whether there is any evidence for the hypothesis that improvements in aerobic fitness will have a larger positive impact on tasks that entail executive control than on tasks involving little if any executive control processing.

Another hypothesis concerning the relation between aerobic fitness and cognition appears, at first glance, to be similar to the current proposal. Chodzko-Zajko and Moore (1994) suggested that tasks requiring controlled, effortful processing should be more sensitive to fitness differences among older adults than tasks that can be executed via automatic processing. Indeed, there is some support for this hypothesis. Fitness was found to have larger effects on choice than on simple RT tasks (Rikli & Edwards, 1991) and on increasingly difficult Stroop interference conditions than on noninterference versions of color and word naming (Schuler, Chodzko-Zajko, & Tomporowski, 1993). However, a number of other studies have failed to find larger fitness effects for tasks that have been assumed to require more effortful (and less automatic) processing (Lupinacci, Rikli, Jones, & Ross, 1993; Toole, Park, & Al-Ameer, 1993). Therefore, at best the support is mixed for Chodzko-Zajko and Moore's (1994) proposal that tasks requiring controlled, effortful processing should be more sensitive to fitness differences among older adults than tasks that can be executed via automatic processing.

It might be argued that this lack of consistency in the relation between fitness and degree of effortful processing belies the hypothesis that improvements in aerobic fitness will have a larger positive impact on tasks that entail executive control than on tasks involving little if any executive control processing. However, the authors argue that tasks requiring effortful processing do not necessarily entail executive control processes. Indeed, many of the experiments that failed to find evidence in favor of Chodzko-Zajko and Moore's (1994) effortful processing hypothesis contrasted fitness effects on simple and choice RT tasks. Whereas choice RT tasks may be more effortful than simple RT tasks, neither of these tasks entails any substantial executive control functions.

Of course, the crucial question is whether there is any evidence in support of the proposed executive control/aerobic fitness hypothesis? The authors believe that there is some, albeit limited, evidence for this proposal. For example, Hawkins et al. (1992) found that subjects trained in water aerobics showed a significantly larger improvement in dual-task performance than a nonexercise control group. Interestingly, both groups showed similar performance improvements in the separate auditory and visual discrimination tasks. Given the obvious coordination and scheduling requirements of successful dual-task performance, as well as previous studies that have found enhanced activation of the prefrontal and frontal regions of the cortex when two tasks are performed together, but not when they are performed separately (D'Esposito et al., 1995), these results are clearly consistent with the executive control hypothesis. Similar results were found in a cross-sectional study that compared fitness effects for a simple RT and a three-choice RT task requiring that subjects continually to remap the stimulus-response relations on the basis of precue information (Abourezk & Toole, 1995). Such a choice RT task clearly requires the executive control processes of planning and coordination as well as the inhibition of previously used response mappings. Consistent with the executive control hypothesis, subjects performed equivalently on the simple RT task, whereas highly fit older adults significantly outperformed their more poorly fit peers on the three-choice task (Emery et al., 1998; Moul et al., 1995; van Boxtel et al., 1997).

A small but growing body of studies is consistent with the hypothesis that improvements in aerobic fitness will have a larger positive impact on tasks that entail executive control than on tasks that involve little if any executive control processing. However, there is an important caveat. None of the reviewed studies were designed explicitly to contrast fitness effects on the executive and nonexecutive processing of older adults.

CURRENT STUDY

In an effort to test the executive control/fitness hypothesis, sedentary but healthy older adults were trained for a period of 6 months with either an aerobic (walking) or anaerobic (toning and stretching control) exercise protocol. Each of the subjects was tested in a variety of attention, memory, and perceptual speed tasks including task switching, response compatibility, stopping, Rey auditory verbal learning (Rey AVL), spatial attention, visual search, N-back spatial and verbal, face recognition, forward and backward digit span, digit–digit and digit–symbol match, self-ordered pointing, and pursuit rotor. These particular behavioral tasks were chosen because components of a subset of these tasks have been shown, through human lesion, neuroimaging, or animal studies, to entail executive control processes and to be supported, in large part, by the frontal or prefrontal regions of the brain.

Components of the first nine tasks listed in Table 6.3 (stopping, task switching, response compatibility, Rey AVL, N-back verbal, N-back spatial and backward digit span, self-ordered pointing and face recognition) have been shown to entail executive processes mediated in the frontal and prefrontal regions of the cortex. These components are as follows: (a) in the stopping paradigm, the stop signal RT (SSRT), which provides a measure of the ability to abort a preprogrammed action; (b) in the task switching paradigm, the difference in RT between the trials during which subjects switch to a new task and the trials during which subjects continue to perform the same task, which provides a measure of task set reconfiguration and inhibition; (c) in the response compatibility paradigm, the difference between RT on the compatible and incompatible trials, which provides a measure of the ability to ignore task-irrelevant stimuli; (d) in the Rey AVLT paradigm, the number of words recalled in list A after the presentation of list B and the ability to recall a list of 15 words in their initially presented order; (e) in the N-back and backward-digit span paradigms, the ability to update and maintain multiple items in working memory, particularly at longer lags; (f) in the self-ordered pointing task, the ability to keep track of and update the items selected during previous trials; and (g) in the face recognition paradigm, the ability to encode and retrieve novel faces successfully.

Performance on the other six paradigms (forward digit span, pursuit rotor, spatial attention, visual search, and digit–digit and digit–symbol), as well as the other components of the stopping (simple RT and choice RT components), task switching (nonswitch trial RT), response compatibility (compatible trial RT), Rey AVL (number of words recalled from list A on the first several trials), and N-back verbal and spatial working memory (1-back) measures (Braver et al., 1997) appear to depend less on executive control processes and the frontal regions of the cortex, and therefore would not be expected to benefit substantially from improvements in aerobic fitness.

In summary, the authors predicted selective cognitive benefits with improved aerobic fitness, and more specifically, benefits in those tasks entail executive control processes supported, in part, by frontal and prefrontal regions of the brain.

METHOD

Subjects

Sedentary older adults (60 to 75 years of age) were recruited to participate in a 6-month randomized exercise intervention. Altogether, 174 individuals passed the initial screening protocol for participation in the study. The screening criteria are illustrated in Table 6.1. Participants were assigned randomly to one of two treatments: an aerobic activity program (walking) and a nonaerobic stretching and toning program. The toning program participants served as a control group against which changes in neurocognitive functions with improved aerobic fitness were evaluated. Subsequently 50 subjects were dropped from the study because of a decision to withdraw from the exercise training or because of incomplete data on the neurocognitive or cardiorespiratory tests. These subjects were divided equally between the walking and the toning groups and did not differ in any of the demographic characteristics from the subjects who completed the studies.

The study was completed by 124 subjects: 66 (20 men) in the toning group and 58 subjects (13 men) in the walking group. The mean ages of the subjects in the toning and walking groups were 66.0 (SD = 5.3) years and 67.3 (SD = 5.2) years, respectively. The two groups also had similar Kaufman K-Bit IQ composite scores (115.2 for the toning group and 113.9 for the and walking groups). The average attendance in the exercise training classes did not differ between the participants in the toning and walking groups.

Neurocognitive Tests

This section provides a brief description of the tasks used in the cognitive battery. Each of these tasks was administered both before and after the 6-month exercise intervention.

Visual Search Task.   This task examined the subjects' ability to search rapidly through a visual array. Two different search tasks were performed by the subjects. In the *feature search task*, subjects detected a target that differed from distractors by a single feature. That is, subjects searched for an X target among O distractors. Display sizes of 5, 10, and 15 items were used. The target was present on 50% of the trials. Subjects depressed one key on the computer keyboard for a target present response and another key for a target absent response. In the *conjunction*

TABLE 6.1

Inclusion Exclusion Criteria for Subject Acceptance Into the Study

Inclusion	Exclusion
1. Age of 60 to 75 years	Younger than 60 years of age
2. Sedentary (no physical activity in the preceeding 6 months)	Self-reported activity on a regular basis (2 times per week) in the preceeding 6 months
3. Capable of performing exercise	Any physical disability that prohibits mobility (e.g., walking, stretching, etc)
4. Personal physician's examination and consent to participate in exercise intervention	Non-consent of physician
5. Successful completion of graded exercise test without evidence of cardiac abnormalities	Evidence of abnormal cardiac responses during graded exercise testing
6. Initial depression score on the GDS below clinical level	Depression score on the Global Depression Scale indicative of clinical depression
7. No history of neurologic disorders	History of neurologic disorders
8. Corrected (near & far) acuity of 20/40 or better	Corrected (near and far) acuity greater than 20/40
9. Fewer than three errors on the Pfeiffer (1975) Mental Status questionnaire	More than three errors on the Pfeiffer questionnaire

search task subjects detected a small X target among large X and small O distractors. The same display sizes, response keys, and target probabilities used in the feature search task were used also in the conjunction search task

Subjects performed one practice block of 20 trials followed by one 120 trial feature search block and two 120 trial conjunction search blocks in the visual search task. The main dependent variables included mean RT, error rate, and RT search slopes.

Response Compatibility Task. This task examined the subjects' ability to ignore task-irrelevant stimuli. Subjects were presented with three letters in the center of the display. They were instructed to attend and respond to the letter in the middle of the three-letter array. If the letter in the center of the array was an X, they were to press one key on the computer keyboard. If the letter in the center of the array was an S, they were to press another key on the keyboard.

There were two different conditions, each occurring on 50% of the trials. In the response-compatible trials, the center letter was flanked by two of the same letters (an X in the middle surrounded by Xs), whereas in the response-incompatible trials, the center letter was flanked by the other letter (an X in the middle surrounded by Ss). The flankers appeared at a .25° visual angle from the target. Subjects performed one practice block of 20 trials followed by one experimental block of 120 trials. The main dependent variables included mean RT and error rate.

Task Switching Paradigm.   This task examined the subjects' ability to disengage rapidly from one task and switch to another. Subjects performed two different tasks that alternated at every third trial. That is, subjects performed an odd/even numerical judgment task (i.e., is a single-digit number odd or even ?) for two trials followed by performance of a vowel/consonant judgment task for two trials. The task stimuli, a letter and a single-digit number, were presented in a two-by-two matrix centered in the middle of the computer screen. When the letter and digit were located in the upper quadrant of the matrix, subjects performed the odd/even judgment task, when the letter and digit were in the lower two quadrants of the matrix subjects performed the consonant/vowel judgment task. The letter and digit were presented in the matrix in a continuous clockwise direction. Thus, the occurrence of a task switch was predictable.

Each stimulus pair was presented until the subject responded. The next stimulus pair was presented 400 ms after the subject's response. Subjects responded with one of two keys on the computer keyboard for each of the tasks.

Subjects first performed two 30-trial single-task blocks followed by one 30-trial task switching block as practice. The practice blocks then were followed by four 60-trial task-switching blocks. The main dependent variables included mean RT and error rate.

Stopping Paradigm.   This task examined the subjects' ability to abort a preplanned manual response rapidly. There were several different task components in the stopping paradigm; a simple RT (SRT) task, a choice RT (CRT) task, and the stopping task. The subjects first performed the SRT task. They manually responded to the presentation of a 75-dB, 100-Hz tone (200-ms duration). Subjects practiced the SRT task for 20 trials, then performed another 50 experimental trials. They then performed the CRT task, in which they were required to discriminate between two letters, an F and an E. The subjects pressed one key on the computer keyboard when the F appeared and another key when the E appeared. They performed the CRT task for 20 practice trials and 60 experimental trials.

The subjects next performed the stopping task. In this task, they were instructed to perform the CRT task unless a tone was presented, which indicated that the response to the CRT task was to be aborted. The tone was presented on 30% of the trials, which are called stop-signal trials. The tones were presented at one of three delays relative to the presentation of the letter. The stop-signal delays were calculated separately for each subject by subtracting the SRT (i.e., the average SRT to the tone-only block of trials) from the RTs from the 20th, 50th and 80th percentile RTs from the CRT block.

The subjects performed 30 practice trials of the stopping task, then three blocks of 120 trials each. The mean RTs and error rates served as dependent variables for the SRT and CRT tasks. Two dependent variables were calculated in the stopping task. One variable was the probability of responding given the occurrence of a stop signal, P(respond/signal), and the second variable was the stop signal reaction time (SSRT). The P(respond/signal) provides a measure of the subject's ability to inhibit an overt response, whereas the SSRT indicates the speed of the stopping process. The SSRT was estimated from the data using the race model proposed by Logan and Cowan (1984).

Spatial Attention Task. This task examined the subjects' ability to reorient attention rapidly in the visual field. The subject's task was to detect and respond to the appearance of a bright asterisk positioned 4° of visual angle to the left or right of fixation in a box. One of two boxes was cued for 50 ms, followed 100 ms later by the appearance of the asterisk target. The cue correctly predicted the location of the asterisk target in 70% of the trials. Subjects performed one block of 35 practice trials followed by two blocks of 55 experimental trials. The main dependent variables included mean RT and accuracy.

Rey Auditory Verbal Learning Test (AVLT). This task examined a number of components of verbal working memory including recognition and recall of a list of ordered stimuli. The subject's task was to recall as many words as possible from two different word lists. In Trials 1 through 5, subjects were read a list (the A list) of 15 common nouns. Then they were asked to recall as many words as possible, in the correct order, as soon as the experimenter has finished reading the word list. In Trial 6, a new list of words (list B) was read and the subjects were asked to recall as many of the 15 words as possible. In Trial 7, no list was read and the subjects were asked to recall as many words as possible from the A list. Trial 8 took place after a 30-minute delay, at which time the subjects were asked again to recall as many words as possible from list A. Trial 9 was a recognition test. The experimenter read a list of 50 words, and the subject was instructed to say "yes" to any word from list A. Finally, Trial 10 was a temporal order task. The subjects were given a sheet of paper with the 15 words from list A presented in a scrambled order. They then were asked to arrange the words in the order they originally were read to them.

The dependent measures were the accuracy of recall (or recognition) for each of the first nine experimental trials and the correlation between the subject's recall order and the correct order of the words on Trial 10.

Pursuit Rotor Task. This task examined the subjects' ability to learn a complex spatial-temporal pattern. The subject's task was to match the position of a computer-controlled target presented on the computer screen with a cursor controlled via a digitizing tablet. On each trial, the target first moved randomly around the screen for 10 secs followed by a movement pattern that repeated across trials for 10 secs, and finally another 5 secs of random movement. Each subject performed eight blocks of four trials each. The first six blocks included the same repeating pattern (the training blocks), whereas a new repeating pattern replaced the initial repeating pattern in blocks seven and eight (the transfer blocks). The dependent variable was the root mean square (RMS) tracking error.

Self-Ordered Pointing Task.   This task examined the subjects' ability to monitor and maintain in working memory a series of responses to stimuli that were not verbally codeable. The subjects performed blocks of nine trials. On each trial nine abstract objects were presented in the cells of a three-by-three matrix on the computer screen. The subject's task was to point to a new object, by using the numeric keypad, on each of nine sequentially presented trials. The objects were presented in a new position in the matrix on each trial. Each subject performed one practice and four experimental blocks of nine trials each. The dependent variable was the average number of unique items that the subject selected.

Spatial Working Memory (N-Back Spatial).   This paradigm examined the subjects' spatial working memory. During each trial the subjects saw a three-by-three matrix with a black box in one of the cells. In the 1-back trial blocks, the subjects were required to make one response if the box in the present trial was in the same location as the box in the previous trial. In the 2-back trial blocks, the subjects were required to decide whether the location of the box, in the three-by-three matrix, was the same as that two trials ago. They performed one practice block of 50 trials and then two experimental trial blocks of 50 trials, one, for the 1-back version and one for the 2-back version of the task. There were a total of 18 trials during which a match occurred in the 1 and 2-back versions of the task. The dependent variable was the number of trials in which the subjects correctly detected a match.

Verbal Working Memory (N-Back Verbal).   This paradigm examined the subjects' verbal working memory. This task was the verbal analog of the spatial working memory task. Instead of matching spatial positions, the subjects were instructed to determine whether letters matched across either one or two trials. The number of practice and experimental trials and the dependent variables were the same as in the spatial working memory task.

Face Recognition Task.   This task examined the subjects' nonverbal working memory, specifically for human faces. A series of 50 black and white photographs of male faces were presented for 3 secs each on the computer screen. The subjects were instructed to provide a verbal report as to whether each of the faces was pleasant or unpleasant. Immediately after the study block, the subjects were given a recognition test, in which 25 pairs of faces were presented (with one new and one old face in each pair) on the computer screen. The subjects were asked to judge which of the two faces was presented in the previous block. After a period of 1 hr, a delayed recognition task was performed. This task was identical to the immediate recognition task, except that the other 25 initially studied faces were paired with new faces. The subjects judged which of the faces in each pair was the "old" face. The dependent measure was the percentage of correctly recognized stimuli in the immediate and delayed recall tests.

Digit–Digit and Digit–Symbol Tests. These tests examined the subjects' ability to perform perceptual comparisons rapidly. In the digit–digit task, the subjects were presented with two rows of nine single digits such that a digit in the top row was aligned with a digit in the second row. The subjects then were presented with a pair of vertically aligned digits and asked to determine, by consulting the rows of digits at the top of the screen, whether the single pair matched one of the pairs of digits at the top of the screen. The digit–symbol task was similar to the digit–digit task. However, in the digit–symbol task numbers were aligned with symbols from the computer keyboard. The subjects performed one practice block of 10 trials and one experimental block of 50 trials for each of these tasks. The dependent variables were mean RT and accuracy.

Forward and Backward Digit Span. These tests examined the subjects' ability to maintain a number of elements in working memory and to retrieve these elements on command. In the forward digit span test, the subjects were presented with sets of digits from the Wechsler Adult Intelligence Scale, Revised (WAIS-R) IQ test and asked to repeat them back to the experimenter in the order of their initial presentation. The backward digit span test was administered in the same manner as the forward digit span test with the exception that the subjects were to recall the digits in the order opposite their initial presentation.

The main dependent variable for the forward and backward digit span tests was the number of points earned by the subjects, with one point received for each string recalled until the point at which subjects failed to recall both digit sets at a particular numerosity.

Exercise Interventions

Aerobic Exercise Group. The aerobic exercise intervention was designed to influence physical fitness as typified by cardiorespiratory endurance. Basic principles and guidelines for exercise programming were followed, including adequate warm-up and cool-down periods, progressive and gradual increments in exercise duration and energy expenditure, and instruction regarding avoidance of exercise-related injury. With respect to the exercise prescription, the intensity level began at light levels (50–55% VO_2max) and gradually increased to more moderate levels (65–70% VO_2max) by the midpoint of the program. The duration of exercise also was increased gradually beginning at 10 to 15 min per session and increasing by a minute per session until participants were exercising for 40 min per session. The exercise classes were conducted by trained exercise specialists and used brisk walking as the aerobic component. The exercise program was conducted three times a week for 6 months.

Activity sessions were conducted initially on the University of Illinois campus and involved participants walking outdoors on one of two premeasured routes. These routes were less than 5 min from the daily starting point and involved walking along, through, and around either a large university quadrangle or a partially wooded parklike area. As the weather conditions became more severe and precluded outdoor activity for these individuals, the activity sessions were moved to a local indoor shopping mall.

Stretching and Toning Control Group.   This group of individuals met three times per week for 6 months in a large gymnasium. They were led by an experienced exercise leader, and therefore received the same amount of attention as the aerobic training group. The focus of this program was on providing of an organized program of stretching, limbering, and toning for the whole body. Each stretch was constant, controlled, and smooth, and progressions were gradual and steady. All the stretches were within each subject's range of motion, and each was held to the point of slight discomfort. This program emphasized stretches for all the large muscle groups of both the upper and lower body. Each stretch was held for approximately 20 to 30 sec and repeated 5 to 10 times. Each stretching/toning session lasted for approximately 40 min. Each session was preceded and followed by 10 min of warm-up and cool-down exercises.

Assessment of Aerobic Capacity.   Aerobic endurance capacity was determined for all subjects before and after the intervention. Because many older individuals are unable to attain an objective verifiable VO_2max, in which a plateau in VO_2 is demonstrated between two or more work levels (treadmill graded) during progressive graded incremental walking, VO_2 peak was assessed at the point of test termination because of volitional exhaustion. This measure of aerobic endurance capacity was assessed on a motor-driven treadmill by using a modified Balke protocol (American College of Sports Medicine, 1991). The protocol involved walking at a speed of 3 miles per hour with increasing grade increments of 2% every 2 min. Measurements of oxygen uptake, heart rate, and blood pressure were continuously monitored.

Oxygen uptake (VO_2) was measured from expired air samples taken at 30-s intervals until a peak VO_2, the highest VO_2, was attained at the point of test termination because of symptom limitation and/or volitional exhaustion. Heart rate was taken during each work stage through continuous direct 12-lead electrocardiographic monitoring. Blood pressure was measured by auscultation and a sphygmomanometer. A physician and nurse monitored and supervised all aspects of the graded exercise testing.

Subjects also completed the Rockport One-Mile Fitness Walking Test (Kline et al., 1987). This measure estimates cardiorespiratory fitness and maximal oxygen uptake and requires subjects to walk 1 mile as quickly as possible. It correlates highly with direct measures of aerobic capacity and has been cross-validated.

RESULTS

The results section is organized in the following manner. First, analyses on the cardiorespiratory measures are reported, both to establish comparability of measures across the walking and toning groups before the exercise intervention and to examine changes in these measures as a function of the exercise interventions.

Next, analyses on the neurocognitive measures are reported. Given the limited space, only the most relevant dependent measures are reported.

Cardiorespiratory Measures

The cardiorespiratory measures obtained in this study are presented in Table 6.2. As can be seen from is table, improvements were observed on the three measures as a function of exercise, with the exception of the VO_2 max measure for the toning group. Each of the three measures was submitted to mixed-mode analyses of variance (ANOVAs), with exercise group (walking and toning) as the between-subjects factor and session (before and after the 6-month exercise intervention) as the within-subjects factor. Significant main effects for session were obtained for time on the treadmill ($F[1,122] = 14.6; p < .01$) and the Rockport 1-mile walk ($F[1,122] = 87.3; p < .01$) measures. The subjects were able to remain on the treadmill longer and complete the 1-mile walk more quickly after the 6 month exercise program. More importantly however, we obtained significant two-way interactions between session and exercise groups were obtained for the VO_2max ($F[1,122] = 10.3; p < .01$), time on treadmill, ($F[1,122] = 9.1; p < .01$) and the Rockport 1-mile walk ($F[1,122] = 7.3; p < .01$) measures. In each of these cases, cardiorespiratory performance improved to a greater extent for the walking than for the toning group as a function of the exercise program.

The percentage of improvement for the walking group on the VO_2 max, the time on treadmill, and the Rockport 1-mile walk measures was 5.1%, 12.9%, and 9.6%, respectively. The comparable percentage improvement scores for the toning group subjects were −2.8%, 1.7%, and 4.4%, respectively.

TABLE 6.2

Pre- and Posttreatment Values for the Cardiorespiratory Tests Administered to the Walking and Toning Groups

Measure	Walking Group		Toning Group	
	Before	After	Before	After
VO_2max (ml/kg/min)	21.5 (0.60)	22.6 (0.64)	21.8 (0.60)	21.2[a](0.54)
Time on treadmill (min)	11.6 (0.35)	13.1 (0.42)	11.3 (0.40)	11.5[a](0.45)
Rockport 1-mile walk (min)	17.7 (0.25)	16.0 (0.26)	18.0 (0.25)	17.2[a] (0.27)

Note: Standard errors are in parentheses.

[a]Indicates a significant interaction ($p < .01$) between exercise group and session.

VO_2max, maximum oxygen consumption.

Neurocognitive Tasks and Measures

The dependent measures in each of the neurocognitive tasks are presented in Table 6.3. The measures for each of the tasks in the neurocognitive battery were submitted to mixed-mode ANOVAs, with exercise group as the between-subjects factor and session as the within-subjects factor. Given the number of paradigms and measures, the focus is specifically on the significant group x session interactions, that is, those interactions that indicate whether the magnitude of performance improvement was larger for the walking or for the toning group subjects as a function of the exercise intervention.

As indicated previously, it was predicted that tasks requiring executive control processes and specifically those processes supported in part by frontal and prefrontal regions of the brain, would be most sensitive to improvements in aerobic capacity. This is, in large part, what was found in the ANOVAs performed with the attentional tasks, and to a lesser extent with the memory tasks.

Stop Signal Paradigm. As indicated in Table 6.3 several different subtasks were included in the stopping paradigm: an SRT task, a CRT task, and the stopping task. It was predicted that differential performance improvements for the two groups would be found for the stopping task, given the need to inhibit inappropriate preprogrammed actions in this task. Indeed, this is what was found. A 17% decrease in SSRT was observed for the walking group after the 6-month exercise program. On the other hand, the toning group subjects showed a 4% increase in SSRT after 6 months ($F[1,122] = 24.7; p < .01$). Also consistent with predictions, no group differences were found on the SRT and CRT tasks, which do not require executive control processes, over the course of the exercise intervention.

Task-Switching Paradigm. As illustrated in Table 6.3, large switch costs (i.e., switch RT – nonswitch RT) were obtained for both groups of subjects, consistent with the previous literature. However, whereas the switch costs declined from 906 to 474 ms for the walking group over the course of the 6-month exercise intervention, the switch costs actually increased for the toning group from 681 to 806 ms. This difference was reflected in the significant three-way interaction among the group, session, and trial type (i.e., switch versus nonswitch trials; $F[1,122] = 12.1$; $p < .01$). Thus, consistent with the selective improvement hypothesis, increases in aerobic fitness had a beneficial effect on those aspects of the task that entailed executive control processes, including inhibition and task set reconfiguration, which are subserved, in part, by prefrontal regions of the brain. Also consistent with the hypothesis, performance on the nonswitch trials, which did not require executive control processes, showed equivalent improvements for both groups.

Response Compatibility Task. A significant three-way interaction among group, session, and compatibility was obtained for the response compatibility task ($F[1,122] = 4.1; p < .05$). Subjects in the walking group reduced their compatibility effect (i.e., response-incompatible RT – response-compatible RT) from 79 ms in the session before exercise training to 25 ms in the session after the 6-month

TABLE 6.3

**Pre- and Posttreatment Values for the Neurocognitive Tests
Administered to the Walking and Toning Groups**

Task Name	Walking Group		Toning Group	
	Before	After	Before	After
Stopping task				
Simple RT (to the tone)	351 (15.1)	361 (13.2)	356 (12.0)	361 (12.6)
Accuracy (for simple RT task)	93.7 (1.9)	94.2 (1.3)	94.3 (1.3)	95.6 (1.1)
Choice RT (to the letters)	664 (16.6)	640 (17.4)	656 (19.1)	635 (18.5)
Accuracy (for choice RT)	90.9 (3.4)	93.1(2.4)	93.6 (1.9)	92.3 (2.8)
Stop signal RT (SSRT)	256 (8.6)	213 (5.4)	249 (6.4)	260[a] (7.5)
P(respond/stop signal) at the intermediate delay	40 (0.02)	39 (0.03)	38 (0.03)	36 (.03)
Task-switching paradigm				
RT (nonswitch trials)	1,590 (71.6)	1,484 (84.8)	1,560 (64.1)	1,457 (81.7)
RT (switch trials)	2,496 (112.9)	1,959 (107.1)	2,241 (96.7)	2,263[a] (117.2)
Accuracy (nonswitch trials)	96.7 (0.85)	97.3 (0.73)	98.5 (0.72)	96.6 (.77)
Accuracy (switch trials)	95.3 (0.87)	95.8 (0.83)	96.4 (0.79)	95.6 (.91)
Response-compatibility task				
RT (compatible trials)	797 (34.7)	701 (23.9)	796 (37.1)	698 (21.9)
RT (incompatible trials)	876 (40.5)	726 (25.4)	868 (35.7)	771[a] (26.7)
Accuracy (compatible trials)	96.7 (0.25)	99.0 (0.17)	98.5 (0.24)	98.6 (0.38)
Accuracy (incompatible trials)	96.8 (1.1)	97.7 (0.87)	97.8 (0.68)	97.7 (0.52)

TABLE 6.3 (*continued*)

Task Name	Walking Group		Toning Group	
	Before	*After*	*Before*	*After*
Rey Auditory Verbal Learning Test				
Trial 1: Number of words recalled from list A	6.1 (0.21)	7.3 (0.21)	5.9 (0.19)	6.9 (0.22)
Trial 5: Number of words recalled from list A	11.8 (0.28)	12.6 (0.26)	11.7 (0.26)	12.6 (0.29)
Trial 6: Number of words recalled from list B	5.5 (0.25)	5.6 (0.25)	5.4 (0.25)	5.6 (0.29)
Trial 7: Number of words recalled from list A	9.3 (0.42)	11.4 (0.37)	9.6 (0.36)	10.3[a] (0.35)
Trial 8: Number of words recalled after 30 min delay	9.7 (0.40)	11.4 (0.38)	9.7 (0.38)	10.9 (0.39)
Trial 9: Number of words recognized	13.4 (0.21)	14.2 (0.15)	13.8 (0.18)	13.9[a] (0.13)
Trial 10: Temporal order ranking-correlation measure	0.60 (0.03)	0.70 (0.03)	0.66 (0.03)	.67[a] (0.03)
N-back verbal				
N-back 1	16.8 (0.21)	17.3 (0.12)	17.1 (.20)	17.4 (0.14)
N-back 2	12.8 (0.47)	14.0 (0.43)	13.5 (0.43)	14.1 (0.44)
N-back spatial				
N-back 1	15.9 (0.36)	16.6 (0.30)	16.9 (0.18)	17.1 (0.21)
N-back 2	11.4 (0.58)	13.8 (0.62)	12.6 (0.44)	14.5 (0.35)
Self-ordered pointing				
Number of unique items	6.1 (0.20)	5.9 (0.15)	6.2 (0.16)	6.1 (0.17)
Face Recognition				
Immediate recognition (% correct)	77.0 (1.9)	80.2 (2.5)	80.4 (1.7)	78.7 (3.0)

TABLE 6.3. (*continued*)

Task Name	Walking Group		Toning Group	
	Before	After	Before	After
Delayed recognition (% correct)	72.1 (1.8)	75.1 (2.4)	74.7 (1.4)	71.4 (2.9)
Backward digit span	6.9 (0.26)	7.2 (0.24)	7.0 (0.26)	7.1 (.26)
Forward digit span	8.2 (0.26)	8.0 (0.26)	8.4 (0.28)	8.4 (.26)
Pursuit rotor task				
Trial 1: RMS for repeating pattern	37.9 (2.0)	32.9 (1.8)	35.6 (1.5)	32.7 (1.4)
Trial 6: RMS for repeating pattern	29.4 (1.2)	28.1 (0.53)	30.2 (1.3)	27.1 (0.51)
Trial 7: RMS for new pattern	29.8 (0.55)	29.5 (0.61)	30.8 (0.56)	30.3 (1.1)
Trial 8: RMS for new pattern	29.4 (0.53)	28.1 (0.39)	29.7 (0.50)	29.1 (0.69)
Spatial attention task				
RT (valid trials)	463 (8.4)	447 (7.7)	468 (8.9)	460 (9.0)
RT (invalid trials)	514 (13.0)	502 (9.8)	522 (11.1)	512 (10.8)
Accuracy (valid trials)	96.9 (0.42)	98.6 (0.32)	98.7 (0.24)	98.4 (0.29)
Accuracy (invalid trials)	92.1 (0.98)	95.6 (0.97)	95.9 (0.91)	95.7 (0.85)
Visual search				
Feature search RT slope	5.2/22.8 (0.38/1.3)	2.8/19.2 (0.36/1.4)	4.6/19.1 (0.50/1.6)	2.9/19.9 (0.31/1.1)[b]
Conjunction search RT slope	7.3/45.2 (0.56/2.4)	7.5/38.1 (0.48/2.1)	7.3/42.1 (0.55/2.6)	7.4/37.6 (0.53/2.1)
Feature search mean accuracy	93.3/99.2 (0.52/0.47)	96.1/99.4 (0.49/0.38)	94.3/98.6 (0.53/0.46)	95.5/98.4 (0.41/0.33)
Conjunction search mean accuracy	94.1/99.1 (0.55/0.40)	97.6/99.4 (0.46/0.39)	96.1/99.4 (0.50/0.45)	96.9/99.1 (0.47/0.39)
Digit–digit match				
RT	839 (16.4)	777 (18.2)	851 (20.1)	804 (18.6)

TABLE 6.3 (continued)

Task Name	Walking Group		Toning Group	
	Before	After	Before	After
Accuracy	97.6 (0.33)	98.4 (0.26)	97.5 (0.38)	97.9 (.34)
Digit–symbol match				
RT	1,907 (53.5)	1,870 (49.1)	1,905 (55.6)	1,884 (61.7)
Accuracy	96.3 (0.49)	96.5 (0.51)	96.1 (0.53)	95.7 (0.78)

Note: Standard errors are in parentheses; reaction times (RT) are in ms; and accuracies are in percentage correct.

[a]Indicates a significant interaction (p<.01) between exercise group and session.

[b]For visual search, the first number is the slope in ms (or SD) for the target present trials, and the second number is the slope (or SD) for the target absent trials.

exercise intervention. However, the response compatibility effect was relatively constant for the subjects in the toning group (i.e., 72 ms and 73 ms in the first and second sessions, respectively). The magnitude of the response compatibility appeared to reflect the subjects' ability to ignore task-irrelevant information successfully, with smaller response compatibility effects being associated with a more successful focus on the task-relevant information. Findings show that the ability to ignore or inhibit conflicting information in response compatibility and other similar tasks successfully is associated with activation of varied prefrontal regions such as the DLPFC and the anterior cingulate (Taylor et al., 1994). Thus, the selective improvement in the response compatibility effect for the aerobic (walking) group is consistent with the study hypothesis concerning executive control processes, aging, and aerobic fitness.

REY Auditory Verbal Learning Test. As indicated in Table 6.3 a number of different measures were obtained from the Rey AVLT, with these measures reflecting different components of memory processes. Given the executive control/fitness hypothesis, the authors predicted that improvements in aerobic fitness would be associated with improvements in performance for a subset of these measures that reflected executive control processes including inhibition (i.e., Trial 7: number of words recalled from list A after presentation of list B) and maintenance of temporal order information (Trial 10: temporal order ranking). Indeed, significant group x session interactions favoring the aerobically trained subjects were found for both the Trial 7 ($F[1,122]=9.9$; $p<.01$) and Trial 10 measures ($F[1,122]=8.5$; $p<.01$). Interestingly, a significant two-way interaction between session and group also was obtained for the Trial 9 measure, which presumably reflects the encoding and retrieval processes underlying recognition ($F[1,122]=5.4$; $p<.05$). Although recognition memory usually is not associated with executive control, the added burden of ensuring that the recognized words were from list A may have been sufficient to invoke inhibitory processes (i.e., to inhibit or suppress the word representations from list B), thereby benefitting recognition memory performance from improved aerobic fitness.

N-back Verbal and Spatial Tasks.   Ample neuroimaging evidence exists to indicate that constantly maintaining and updating working memory as well as determining whether previously encoded items match current items, especially with higher lags, engages a variety of executive control processes subserved by regions in the frontal and prefrontal areas of the cortex (Braver et al., 1997). Therefore, the authors expected to observe significant group x session interactions for accuracy measures in these tasks, and in particular for the lag-2 measures. Although larger improvements in performance were observed for lag 2 measures in the walking than in the toning group for both the verbal and spatial tasks, neither of these effects were significant ($p > .10$).

Self-Ordered Pointing Task.   Given that this task entails the ability to monitor and maintain in working memory a series of responses to stimuli that are not verbally codeable as well as previous demonstrations of the integral role of mid- and posterior dorsolateral cortex in successful performance of the task (Petrides, Alivisatos, Evans, & Meyer, 1993), the authors expected to observe a larger improvement in performance for the aerobically trained subjects. However, as shown in Table 6.3 neither of the subject groups showed any improvement in performance over the course of the intervention.

Face Recognition Task.   As can be seen in Table 6.3, subjects in the walking group showed improvements in both the immediate and delayed recognition measures of the task. This is in contrast to declines in recognition performance for the toning group's subjects over the period of the exercise intervention. However none of the relevant interactions were significant ($p > .10$). Thus, although the trends in performance were in the predicted direction on the face recognition test, they did not achieve conventional levels of statistical significance.

Backward and Forward Digit Span Tasks.   The authors predicted a group x session interaction favoring the aerobically trained adults for the backward, but not for the forward digit span tasks, because whereas the forward digit span task provides a measure of short term memory, it does not require the temporal order transformations observed with the backward memory digit span task that forms one hallmark of executive control processes (Baddeley, 1992). Neither the main effect of session nor the interaction of session and group attained significance.

Pursuit Rotor Task.   Given the minimal level of executive control required for the pursuit rotor task, whether with repeating or new patterns, the authors predicted that any performance differences observed over the course of the exercise intervention would be the same for both the walking and toning groups. Indeed, a main effect was obtained ($F[1,122] = 5.3$; $p < .01$) for session. However, as predicted, the session x group interaction was not significant ($F < 1$).

Spatial Attention Task.   Given that this task appears to rely mainly on parietal cortex and midbrain structures rather than the frontal lobes (Posner & Petersen, 1990) and entails little or no executive control, the authors predicted equivalent performance changes for the walking and toning group subjects over the course of the 6-month exercise intervention. The results were consistent with the predictions. The amount of improvement in the speed of responding did not differ between the toning and walking groups ($F < 1$).

Visual Search Task.   Previous studies have suggested that striate, extrastriate, parietal, and temporal cortices play an important role in visual search for both feature and conjunction targets (Posner & Gilbert, 1999). Therefore, the authors' expectation was that although improvements in search might be observed, as a function of practice over the course of the 6-month exercise intervention, such improvements would be found for both the toning and walking groups. As indicated in Table 6.3, a main effect was observed for session ($F[1,122] = 12.5$; $p < .01$). The RTs were faster in the session after the exercise intervention. However, group did not interact with session in the two- or three-way interactions ($p > .50$). Thus, consistent with the authors' predictions, improvements in performance occurred for both aerobically and nonaerobically trained adults.

Digit–Digit and Digit–Symbol Tasks.   These tasks entail simple comparison and response processes that are unlikely to involve more than minimal executive control (Salthouse 1996). Therefore, the authors predicted that changes in performance on these two perceptual comparison tasks would occur for both the toning and walking groups. Improvements in response speed were observed in both tasks after the 6-month exercise interventions (digit–digit, $F[1,122] = 6.7$; $p < .05$, digit–symbol, ($F[1,122] = 4.6$; $p < .05$). However, neither of these effects interacted with group ($p > .35$).

Regression Analyses

In general, the results were consistent with the predictions. That is, those tasks (and specific components of tasks) that involve executive control processes and are supported in part by frontal and prefrontal regions of the brain showed improvements in performance for the walking but not for the toning group subjects. This was the case for components of the task-switching, stopping, response-compatibility, and Rey AVLT tasks. On the other hand, tasks that involve little or no executive control were not expected to benefit from aerobic training. Consistent with this prediction, equivalent performance benefits, likely the result of practice effects, were found for the walking and toning groups for the pursuit rotor, spatial attention, visual search, digit–digit, and digit–symbol tasks.

However, the predictions were not confirmed for all of the tasks. Selective improvements also were expected for a variety of working memory tasks including the N-back tasks, self-ordered pointing, digit span, and face recognition given the role of executive control processes and regions of prefrontal and frontal cortex in these paradigms. Although there were performance improvement trends in the predicted direction for a variety of these tasks (or task components), they did not attain conventional levels of statistical significance.

There certainly are a multitude of reasons why the predictions might not have been confirmed in these particular tasks. However, one possibility that could be examined with the current data set concerned the relation between the magnitude of improvement in aerobic fitness and the potential improvement in the performance of these tasks. Inspection of the distributions showing the percentage of VO_2 max change, as a function of the 6-month exercise programs indicated a fairly substantial overlap in these distributions for the walking and toning groups. The percentage of VO_2 change ranged from −30% to +45 % across the groups.

Therefore, given the substantial overlap in the distributions showing the percentage of VO_2 max change for the walking and toning group subjects, it was decided to examine the relation between the change in aerobic fitness, as indexed by the VO_2max measure, and performance changes in our battery of tasks over the course of the 6 month exercise interventions. That is, group distinctions (i.e. toning or walking group assignment and training) were ignored and the aerobic and performance change measures were analyzed as continuous variables. To this end, a series of hierarchical regression analyses was performed, in which the VO_2max score obtained before the exercise intervention (i.e., to control for absolute level of aerobic efficiency) was entered on the first step, with the VO_2max postintervention score allowed to enter thereafter as an additional predictor if it accounted for significant residual variance in the performance change score. The VO_2 max postintervention predictor accounted for significant residual variance in (a) the SSRT measure in the stopping task (15.0%), (b) the switch cost measure in the task switching paradigm (12.2%), (c) the response interference measure in the response-compatibility task (14.4%), (d) Trial 7 (9.1%), Trial 9 (8.4%), and Trial 10 (10.3%) measures in the Rey AVLT task, (e) the 2-back measures in the verbal (7.7%) and spatial (6.4%) N-back tasks, (f) the backward-digit-span measure (5.0%), and (g) the immediate (6.9%) and delayed recall (5.8%) measures in the face recognition task.

The regression analyses therefore indicated a significant relation between improvement in VO_2max over the course of the 6-month exercise interventions and improved performance of both the tasks in which significant group x session interactions were obtained in the ANOVA as well as the memory tasks (i.e., N-back, face recognition, backward digit span), which the authors had predicted would be sensitive to improved aerobic fitness. The only exception was the self-ordered pointing task, for which a significant relation was not found between VO_2max change and performance change over the course of the 6 month exercise intervention.

DISCUSSION

The current study examined the relation between improved aerobic fitness engendered by a 6-month fitness training program and improvements in cognitive function on a variety of attentional and memory paradigms. More specifically, the study tested the hypothesis that improvements in performance would be observed to the extent that the task components relied on executive control processes such as monitoring, scheduling, planning, inhibition, and working memory, that is, those processes subserved by the frontal and prefrontal regions of the brain (Shallice,

1994). To that end, tasks were included in the computerized cognitive assessment battery that required executive control processing as well as tasks and task components that did not require such processes.

The initial analyses, which involved ANOVAs on performance measures as a function of subject group and session, provided strong support for the study hypothesis with the attention tasks. Improvements in performance were found for the walking but not for the toning group in the predicted components of the response-compatibility, stopping, and task-switching paradigms, but not in components of these tasks that did not entail executive control. Additionally, as predicted, selective benefits were not found for the performance of the aerobically trained subjects on the spatial attention, visual search, digit–digit, and digit–symbol tasks. These tasks do not entail executive control processes and are subserved, in large part, by nonfrontal regions of the cortex.

However, the initial analyses of the memory paradigms were less encouraging. Although, as predicted, selective performance benefits were found on a subset of the measures of the Rey AVLT that involve executive control processes, no such benefits were observed for the N-back, face-recognition, backward-digit-span, and self-ordered pointing tasks, although there were trends in the predicted direction for a subset of these paradigms. However, the hierarchical regressions did, for the most part, provide data more supportive of the selective improvement hypothesis. When group distinctions (i.e., distinctions between the walking and toning training groups) were ignored, and the focus was instead on changes in aerobic fitness across groups, it was found after baseline VO_2max were first taken into account, differences that the aerobic fitness changes accounted for significant residual variance, in the performance on a number of memory tasks predicted to be sensitive to fitness effects. Indeed, the same types of dissociations were observed between task components with substantial executive control demands, as compared with task components with little demand for executive control processes in a number of analyses involving session x group interactions for the attention tasks. For example, for the spatial and verbal N-back tasks, significant VO_2max performance relations were found in the 2-back but not in the 1-back conditions. Such a finding is consistent with the authors' predictions given previous reports of graded patterns of frontal lobe activation with larger n-back lags (Braver et al., 1997). Similar dissociations were found between the forward and backward digit-span tasks.

One interesting issue concerns the relative robustness of the fitness–performance relation for the attention and memory tasks. Why could the fitness–performance relation be discerned in both the group comparisons and the regression analyses for the attention but not for the memory tasks? Although a more definitive answer to this question must await additional research, there are several possibilities to consider. First, previous meta-analyses of mnemonic training effectiveness with older adults have indicated larger training benefits with younger seniors (Verhaeghen, Marcoen, & Goossens, 1992). Therefore, it is conceivable that any aerobic fitness benefits might also be larger for younger than older participants. To examine this issue, a median split on age was performed, and the ANOVAs were recomputed. The results were quite clear. No significant interactions of age were found with group and session for any of the attention or memory tasks. Therefore, at least in the current case, it appears that aerobic fitness benefits (or the lack thereof) did not depend on the age of the participants.

Another possible reason for the differential strength in the relation between fitness and performance for the attention and memory tasks might be the subset of executive control processes and their underlying neuroanatomical substrates, which are involved in the tasks. For example, although all of the memory tasks required both encoding and retrieval operations, the attentional tasks were more focused on mapping simple stimuli to responses while ignoring task-irrelevant stimuli and actions. Although the executive control processes that support performance in both the attention and memory tasks all have been shown to be influenced by aging, they also are somewhat distinct, thereby making it conceivable that they might be differentially sensitive to changes in aerobic fitness. One way to address further the issue of differential sensitivity of different executive control processes to aerobic fitness would be to include tasks with multiple components (e.g., working memory tasks in task-switching and stopping paradigms) in assessment batteries.

A somewhat related issue concerns the dependent measures and subject strategies used in the attention and memory tasks. The attention tasks stressed rapid and accurate performance, whereas the memory tasks stressed accurate performance. Indeed, the strongest fitness effects were obtained with the reaction time and not the accuracy measures. Therefore, it is possible that aerobic fitness benefits might be observed more readily under speed stress conditions. Such a possibility could be examined in future studies by varying the speed/accuracy emphasis on a subset of the assessment tasks. In any event, there are clearly a number of hypotheses concerning the strength of fitness–performance relations that merit additional study.

In the initial discussion on the aging, fitness, and cognition literature, another hypothesis concerning the relation between these factors was briefly described. Chodzko-Zajko and Moore (1994) suggested that tasks requiring controlled, effortful processing should be more sensitive to fitness differences among older adults than tasks that can be executed via automatic processing. Can this hypothesis also account for the pattern observed in the current study? Given the usual association of more effortful processing with slower performance, the answer is no. The stopping paradigm can be considered as an example (Table 6.3). The executive control/fitness hypothesis predicted that aerobic fitness benefits would be found only in the measure of stopping performance (i.e., SSRT), and not in measures of performance in the SRT and CRT components, although these components took longer to perform than aborting a preprogrammed response. On the other hand, the effortful processing hypothesis predicted stronger fitness/performance benefits for the more difficult tasks (i.e., tasks performed more slowly). Clearly, the data are consistent with the executive control hypothesis and not the effortful processing hypothesis. Of course, another problem for the effortful processing hypothesis involves mapping levels of effortfulness to different task components (i.e., how is a scale of effortfulness defined?).

Throughout this article, an attempt was made to draw parallels between age-related changes in cognition, neuroanatomy, and neurophysiology. Indeed, there is strong support both from human lesion and neuroimaging studies for the link between executive control and frontal lobe processes (Corbetta et al., 1991; D'Esposito et al., 1995; Petrides et al., 1993). Disproportionate age-related changes also have been reported in prefrontal and frontal lobe function and structure and in the executive processes these regions support (Raz et al., in press; West, 1996). However, thus far, no attempts have been made to examine the relation among aerobic fitness, cognition, and brain function and structure. This is not surprising given the fairly recent developments in functional neuroimaging and the complexity of conducting fitness interventions with older adults. However, given the increasing knowledge concerning the relation between fitness and brain function and structure gained in animal research (Jones et al., 1998; Neeper et al., 1995) along with the development of nonintrusive neuroimaging techniques such as fMRI (D'Esposito et al., 1999), the time is now ripe to examine fitness effects on both the brain and mind.

Another important issue that remains to be addressed is the degree to which the cognitive changes resulting from improved aerobic fitness influence activities of daily living for older adults. Clearly, many of the tasks that we perform on a day-to-day basis, such as driving, planning and cooking a meal, and working in a busy office, entail substantial executive control demands. Therefore, it certainly is conceivable that the cognitive improvements measured on relatively simple laboratory tasks such as those used in the current study will translate into improvements in the performance of everyday tasks. However, a more systematic examination of this issue is required to confirm such a speculation.

ACKNOWLEDGMENTS

This research was supported by grants AG12113 and AG14966 from the National Institute on Aging. A subset of the data presented here was presented originally in Kramer et al. (1999). The authors thank Heidi Prioux and Meredith Minear for their assistance in running subjects. They also thank Wendy Rogers and an anonymous reviewer for their helpful comments on a previous version of this chapter. Requests for reprints can be sent to Arthur F. Kramer, Beckman Institute, University of Illinois, 405 North Mathews Avenue, Urbana, Illinois, 61801 or akramer@s.psych.uiuc.edu.

REFERENCES

Abourezk, T., & Toole, T. (1995). Effect of task complexity on the relationship between physical fitness and reaction time in older women. *Journal of Aging and Physical Activity, 3*, 251–260.

Allport, A., Styles, E., & Hsieh, S. (1994). Shifting intentional set: Exploring the dynamic control of tasks. In C. Umilta & M. Moscovitch (Eds.), *Attention and performance XV* (pp. 421–452). Cambridge, MA: MIT Press.

American College of Sports Medicine. (1991). *ACSM's Gridlines for Exercise Testing and Prescription* (4th ed.). Baltimore: Williams and Wilkins.

Azari, N., Rapport, S., Salerno, J., Grady, C., Gonzales-Aviles, A., Schapiro, M., & Horwitz, B. (1992). Intergenerational correlations of resting cerebral glucose metabolism in old and young women. *Brain Research, 552*, 279–290.

Baddeley, A. (1992). Working memory. *Science, 255*, 556–559.

Barry, A. J., Steinmetz, J. R., Page, H. E., & Rodahl, K. (1966). The effects of physical conditioning on older individuals. II. Motor performance and cognitive function. *Journal of Gerontology, 21*, 192–199.

Blumenthal, J. A., & Madden, D. J. (1988). Effects of aerobic exercise training, age, and physical fitness on memory search performance. *Psychology and Aging, 3,* 280–285.

Blumenthal, J. A., Emery, C. F., Madden, D. J., Schniebolk, S., Walsh-Riddle, M., George, L. K., McKee, D. C., Higginbotham, M. B., Cobb, F. R., & Coleman, R. E. (1991). Long-term effects of exercise on psychological functioning in older men and women. *Journal of Gerontology: Psychological Sciences, 46,* 352–361.

Boutcher, S. H., & Landers, D. M. (1988). The effects of vigorous exercise on anxiety, heart rate, and alpha activity of runners and nonrunners. *Psychophysiology, 25,* 696–702.

Braver, T. S., Cohen, J. D., Nystrom, L. E., Jonides, J., Smith, E. S., & Noll, D. C. (1997). A parametric study of prefrontal cortex involvement in human working memory. *Neuroimage, 5,* 49–62.

Brink, J., & McDowd, J. (1999). Aging and selective attention: An issue of complexity or multiple mechanisms? *Journal of Gerontology: Psychological Sciences, 54,* 30–33.

Bunce, D. J., Barrowclough, A., & Morris, I. (1996). The moderating influence of physical fitness on age gradients in vigilance and serial choice responding. *Psychology and Aging, 11,* 671–682.

Chodzko-Zajko, W., & Moore, K. A. (1994). Physical fitness and cognitive function in aging. *Exercise and Sport Science Reviews, 22,* 195–220.

Clarkson-Smith, L., & Hartley, A. (1990). Structural equation models of relationships between exercise and cognitive abilities. *Psychology and Aging, 5,* 437–446.

Cook, N., Albert, M., Berkman, L., Blazer, D., Taylor, J., & Hennekens, C. (1995). Interrelationships if peak expiratory flow rate with physical and cognitive function in the elderly: MacArthur foundation studies of aging. *Journal of Gerontology: Medical Sciences, 50,* 317–323.

Corbetta, M. (1998). Frontoparietal cortical networks for directing attention and the eye to visual locations: Identical, independent, or overlapping neural systems. *Proceedings of the National Academy of Science, 95,* 831–838.

Corbetta, M., Miezin, F. M., Dobmeyer, S., Shulman, G. L., & Petersen, S. E. (1991). Selective and divided attention during discriminations of shape, color, and speed: Functional anatomy by positron emission tomography. *The Journal of Neuroscience, 11(8),* 2383–2402.

Cotman, C., & Neeper, S. (1996). Activity dependent plasticity and the aging brain. In E. Schneider & J. Rowe (Eds.), *Handbook of the biology of aging* (pp. 284–299). New York: Academic Press.

Daigneault, S., Braun, C., & Whitaker, H. (1992). Early effects of normal aging on perseverative and non-perseverative prefrontal measures. *Developmental Neuropsychology, 8,* 99–114.

D'Esposito, M., Detre, J., Alsop, D., Shin, R., Atlas, S., & Grossman, M. (1995). The neural basis of the central executive system of working memory. *Nature, 378,* 279–281.

D'Esposito, M., Zarahn, E., Aguirre, G. K., & Rypma, B. (1999). The effect of normal aging on the coupling of neural activity to the Bold hemodynamic response. *Neuroimage, 10,* 6–14.

Dustman, R., Emmerson, R, & Shearer. (1994). Physical activity, age, and cognitive neuropsychological function. *Journal of Aging and Physical Activity, 2,* 143–181.

Dustman, R. E., Emmerson, R. Y., Ruhling, R. O., Shearer, D. E., Steinhaus, L. A., Johnson, S. C., Bonekat, H. W., & Shigeoka, J. W. (1990). Age and fitness effects on EEG, ERPs, visual sensitivity, and cognition. *Neurobiology of Aging, 11,* 193–200.

Dustman, R. E., Ruhling, R. O., Russell, E. M., Shearer, D. E., Bonekat, W. , Shigeoka, J. W., Wood, J. S., & Bradford, D. C. (1984). Aerobic exercise training and improved neurophysiological function of older adults. *Neurobiology of Aging, 5,* 35–42.

Elsayed, M., Ismail, A., & Young, R. (1980). Intellectual differences of adult men related to physical fitness before and after an exercise program. *Journal of Gerontology, 35,* 383–387.

Emery, C. F., Schein, R. L., Hauck, E. R., & MacIntyre, N. R. (1998). Psychological and cognitive outcomes of a randomized trial of exercise among patients with chronic obstructive pulmonary disease. *Health Psychology, 17*, 232–240.

Fordyce, D. E., & Farrar, R. P. (1991). Physical activity effects on hippocampal and parietal cholinergic function and spatial learning in F344 rats. *Behavioral Brain Research, 43*, 115–123.

Hartley, A. (1993). Evidence for selective preservation of spatial selective attention in old age. *Psychology and Aging, 8*, 371–379.

Hawkins, H., Kramer, A., & Capaldi, D. (1992). Aging, exercise, and attention. *Psychology and Aging, 7*(4), 643–653.

Hill, R. D., Storandt, M., & Malley, M. (1993). The impact of long-term exercise training on psychological function in older adults. *Journal of Gerontology: Psychological Sciences, 1*, 12–17.

Jones, T. A., Hawrylak, N., Klintsova, A. Y., & Greenough, W. T. (1998). Brain damage, behavior rehabilitation recovery, and brain plasticity. *Mental Retardation and Developmental Disabilities Research Reviews, 4*, 231–237.

Kline, G. M., Pocari, J. P., Hintemeister, R., Freedson, P. S., Ward, A., McCarron, R. F., Ross, J., & Rippe, J. M. (1987). Estimation of Vo_2max from a 1-mile track walk, gender, age, and body weight. *Medical Science, Sports, and Exercise, 19*, 353–359.

Kramer, A. F., Hahn, S., Cohen, N. J., Banich, M. T., McAuley, E., Harrison, C. R., Chason, J., Vakil, E., Bardell, L., & Colcombe, A. (1999a). Aging, fitness, and neurocognitive function. *Nature, 400*, 418–519.

Kramer, A. F., Hahn, S., & Gopher, D. (1999b). Task coordination and aging: Explorations of executive control processes in the task switching paradigm. *Acta Psychologica, 101*, 339–378.

Kramer, A. F., Hahn, S., Irwin, D. E., & Theeuwes, J. (2000). Age differences in the control of looking behavior: Do you know where your eyes have been? *Psychological Science, 11*, 210–217.

Kramer, A. F., Larish, J., Weber, T., & Bardell, L. (1999). Training for executive control: Task coordination strategies and aging. In D. Gopher & A. Koriat (Eds.), *Attention and performance XVII*. Cambridge, MA: MIT Press.

Logan, G. D., & Cowan, W. B. (1984). On the ability to inhibit simple and choice reaction time responses: A model and a method. *Psychological Review, 91*, 295–327.

Lupinacci, N. S., Rikli, R. E., Jones, C. J., & Ross, D. (1993). Age and physical activity effects on reaction time and digit symbol substitution performance in cognitively active adults. *Research Quarterly for Exercise and Sport, 64*, 144–150.

Madden, D. J., Blumenthal, J. A., Allen, P. A., & Emery, C. F. (1989). Improving aerobic capacity in health older adults does not necessarily lead to improved cognitive performance. *Psychology and Aging, 4*, 307–320.

Marchal, G., Rioux, P., Petit-Taboue, M. C., Sette, G., Ttavere, J. M., LePoec, C., Courtheoux, P., Derlon, J. M., & Baron, J. C. (1992). Regional cerebral oxygen consumption, blood flow, and blood volume in healthy human aging. *Archives of Neurology, 49*, 1013–1020.

Meyer, J. S., Kawamura, J., & Terayama, Y. (1994). Cerebral blood flow and metabolism with normal and abnormal aging In M. Albert & J. Knoefel (Eds.), *Clinical neurology of aging* (pp. 214–234). New York: Oxford University Press.

Moul, J., Goldman, B., & Warren, B. (1995). Physical activity and cognitive performance in the older population. *Journal of Aging and Physical Activity, 3*, 135–145.

Neeper, S., Gomez-Pinilla, F., Choi, J., & Cotman, C. (1995). Exercise and brain neurotrophins. *Nature, 373*, 109.

Offenbach, S., Chodzko-Zajko, W., & Ringel, R. (1990). Relationship between physiological status, cognition, and age in adult men. *Bulletin of the Psychonomic Society, 28*, 112–114.

Panton, L. B., Graves, J. E., Pollock, M. L., Hagberg, J. M., & Chen, W. (1990). Effect of aerobic and resistance training on fractionated reaction time and speed of improvement. *Journal of Gerontology: Medical Sciences, 45*, 26–31.

Petrides, M., Alivisatos, B., Evans, A. C., & Meyer, E. (1993). Dissociation of human mid-dorsolateral from posterior dorsolateral frontal cortex in memory processing. *Proceedings of the National Academy of Science, 90*, 873–877.

Posner, M. I., & Gilbert, C. D. (1999). Attention and primary visual cortex. *Proceedings of the National Academy of Science, 96*, 2585–2587.

Posner, M., & Petersen, S. (1990). The attention system of the human brain. *Annual Review of Neuroscience, 13*, 25–42.

Raz, N. (in press). Aging of the brain and its impact on cognitive performance: Integration of structural and functional findings. In F. Craik & T. Salthouse (Eds.), *Handbook of aging and cognition*. New Jersey: Lawrence Erlbaum Associates.

Rikli, R., & Edwards, D. (1991). Effects of a three-year exercise program on motor function and cognitive processing speed in older women. *Research Quarterly for Exercise and Sport, 62*, 61–67.

Rogers, D., & Monsell, S. (1995). Costs of a predictable switch between simple cognitive tasks. *Journal of Experimental Psychology: General, 124*, 207–231.

Rogers, W. A., Bertus, E. L., & Gilbert, D. K. (1994). A dual-task assessment of age differences in automatic process development. *Psychology and Aging, 9*, 398–413.

Salmon, E., Marquet, P., Sandzot, B., Degueldre, C., Lemaire, C., & Franck, G. (1992). Decrease of frontal metabolism demonstrated by positron emission tomography in a population of healthy elderly volunteers. *Acta Neurologica Belqique, 91*, 288–295.

Salthouse, T. (1996). The processing-speed theory of adult age differences in cognition. *Psychological Review, 103*, 403–428.

Schuler, P. B., Chodzko-Zajko, W. J., & Tomporowski, P. (1993). Relationship between physical fitness, age, and attentional capacity. *Sports Medicine Training and Rehabilitation, 4*, 1–6.

Shallice, T. (1994). Multiple levels of control processes. In C. Umilta & M. Moscovitch (Eds.), *Attention and performance XV* (pp. 395–420). Cambridge, MA: MIT Press.

Shimamura, A. P., & Jurica, P. J. (1994). Memory interference effect and aging: Findings from a test of frontal lobe function. *Neuropsychology, 8*, 408–412.

Spirduso, W. W. (1975). Reaction and movement time as a function of age and physical activity level. *Journal of Gerontology, 30*, 18–23.

Spirduso, W. W., & Clifford, P. (1978). Replication of age and physical activity effects on reaction time movement time. *Journal of Gerontology, 33*, 23–30.

Stones, M., & Kozma, A. (1988). Age, exercise and coding performance. *Psychology and Aging, 4*, 190–194.

Taylor, S. F., Kornblum, S., Minoshima, S., Oliver, L. M. (1994). Changes in medial cortical blood flow with a stimulus response compatibility task. *Neuropsychologia, 32*, 249–255.

Toole, T., Park, S., & Al-Ameer, H. (1993). Years of physical activity can affect simple and complex cognitive/motor speed in older adults. In G. E. Stelmach & V. Homberg (Eds.), *Sensorimotor impairment in the elderly* (pp. 427–439). Dordrecht, The Netherlands: Kluwer Academic.

van Boxtel, M., Paas, F., Houx, P., Adam, J., Teeken, J., & Jolles, J. (1997). Aerobic capacity and cognitive performance in a cross-sectional aging study. *Medicine and Science in Sports and Exercise, 10*, 1357–1365.

Verhaeghen, P., Kliegl, R., & Mayr, Y. (1997). Sequential and coordinative complexity in time-accuracy functions for mental arithmetic. *Psychology and Aging, 12*, 555–564.

Verhaeghen, P., Marcoen, A., & Goossens, L. (1992). Improving memory performance in the aged through mnemonic training: A meta-analytic study. *Psychology and Aging, 7,* 242–251.

West, R. (1996). An application of prefrontal cortex function theory to cognitive aging. *Psychological Bulletin, 120,* 272–292.

PART III

HOME HEALTH CARE AND CAREGIVING

7

The State of Telecommunication Technologies to Enhance Older Adults' Access to Health Services

Pamela Whitten
Michigan State University

Health care in the United States has evolved over this century from a service accessed by a small percentage of society to one of the biggest market sectors in our economy. The Healthcare Financing Administration (1999) reported that the nation's total spending for health care will increase from $1.0 trillion in 1996 to $2.1 trillion in 2007. This growth represents an increase in health care costs as a percentage of gross domestic product (GDP) from 13.6% to almost 17%. These national trends in health expenditures are attributed to a boost in underlying demand for medical services resulting from a recent growth in real per capita income accompanied by increasing medical inflation. Patterns in health care growth, however, vary widely across types of services. It is anticipated that hospital growth will fall well short of the increase in physician and professional services as the trend from inpatient to outpatient care continues. The fastest growing segment of health care expenditures is projected to be for drugs because of an increase in the number of actual prescriptions and changes in the size and mix of medications.

As the largest consumer segment in the health industry, elderly Americans will simultaneously have the largest impact on these shifts in health costs and services and be the most impacted by them. Shifts in population composition play an important role in influencing many of the policies and programs implemented in our nation. A report by the U.S. Bureau of the Census (1996) stated that the growth rate of people older than 65 years would be relatively modest through the year 2005. However, when persons born between 1946 and 1964 (the baby-boom generation) begin reaching the age of 65 years, we will witness a rapid increase in the population of older Americans. About one in eight Americans were elderly in 1994. However, it is projected that one in five will be elderly by 2030.

Any attempt to shape public policy around the simple notion that we are about to face a shift toward an older population will miss many important nuances that should also guide decisions regarding health services and allocation. For example, the term "elderly" is commonly used to describe anyone 65 years of age or older. Yet, this heterogeneous group actually is composed of the young old (65–74 years old), the aged (75–84 years old), and the oldest old (85 years and older). Moreover, each of these subgroups has very different health needs. One study documented that 75% of the noninstitionalized elderly between the ages of 65 and 74 years consider their health to be very good as compared with 66% of those older than 75 years. The oldest old require more long-term assistance, with nearly 25% living in a nursing home (U.S. Bureau of the Census, 1996).

In addition to major health needs based on age range, it also is known that the rural elderly have characteristics and needs that differ from those of their urban counterparts. For example, they often are more isolated, have lower incomes, and have greater challenges actually accessing health services because of transportation needs. Rural residents are more likely to put off care as a result of cost or transportation difficulties (Blazer, Landerman, Fillenbaum, & Horner, 1995). Rural residents also are less likely to obtain health care at a hospital. Although the older population is almost universally covered by Medicare and Medicaid in the United States, substantial differences in out-of-pocket expenses mean health care is not equally accessible to all (Blazer et al., 1995).

There are additional demographic characteristics that have an impact on the potential typography of health services. For example, elderly women in the United States outnumber elderly men 3 to 2. Furthermore, because the death of a husband often leads to acute economic changes for the surviving wife, health analysts need to consider changing needs based on economic constraints. In the past, the elderly population has been predominately white. However, the coming decades are expected to produce an elderly population that is more racially and ethnically diverse, with the percentage of nonwhite elderly expected to increase from ten to 20 percent between 1990 and 2050.

As medical innovations provide health care providers with enhanced ways to save lives, we can expect to see the length of chronic illnesses and the subsequent need for supporting services increase dramatically in the coming years. Consequently, communities are now seeking new solutions for accompanying issues of rising costs, improved access to services, and acceptable levels of quality and continuity of care. Technology is emerging as one approach to addressing some of these issues for a variety of reasons. First, the 1990s has evidenced unprecedented developments in the telecommunications industry, which have resulted in a powerful information and telephony infrastructure. This infrastructure is providing increasing access to interactive data and communication services from a variety of settings. Because of these high-speed electronic connections, time, travel, and location for patients and providers take a backseat to timelier, better-coordinated, and more efficient health services. For example, new technologies have made it feasible to bring health care to a patient's geographic location and in effect make house calls.

A second explanation for the explosive growth in the use of technology for health services highlights explosion in the development of new technologies created to deliver services. One example is the creation of standards for interactive video (ITV) equipment. Before the middle of the 1990s, most ITV units operated by proprietary algorithms. This meant that no two brands could talk to each other. With the development of standards (e.g., H.320; H.324), equipment from almost any vendor can "talk to" different brands. Another example of emerging equipment is a class of technology called telemedicine peripheral devices. These are medical devices that can be used on patients over a distance to support the needs of a medical consult or nursing visit. An electronic stethoscope is one example, whereby a patient's heart and breath sounds can be heard synchronously over great distances through telephone lines.

A final reason for the explosion of technologies to deliver medical services involves our society's growing acceptance of technologies as an inevitable part of our lives. Although these innovations are far from being completely diffused, there is increasing evidence that they are making their way into mainstream society (e.g., HCFA legislation to reimburse for telemedicine services). The bottom line is the need to enhance the range of services and access to them while simultaneously lowering overall costs. Telemedicine and telecommunication-based technologies may prove to be one tool that can assist in attaining these goals.

The purpose of this chapter is to look at this important sector of health care for the elderly: the use of technologies to deliver medical services and health education.[1] A deductive approach is used, which begins by providing a foundation of the evolution and effects of technology in the health arena. Next, the discussion focuses specifically on traditional telemedicine, the use of telecommunication technologies to provide health services, and provides an overview of telemedicine applications for the elderly. An in-depth discussion of the newest generation of telemedicine, e-health, follows. Finally, this chapter concludes with representative evaluation issues associated with ehealth that will have direct and indirect impacts on access and quality of care for older adults.

HEALTH CARE AND TECHNOLOGY

Before 1900, the medical establishment had little to offer in the way of treatment and cures because its resources were primarily the physician, his education, and the mysterious tools in his little black bag. Although the costs of obtaining medical care from a physician were relatively cheap, the demand for a physician's services was small because amateurs residing in a community could provide many health services. Typically, a patient's home was the site for treatment and recuperation, and family and neighbors acted as the nursing and support staff. However, the

---

[1] Please note that this chapter simply seeks to overview the state of telecommunication technologies in the field of older adult health care. Readers interested in topics such as relational issues between patients and providers or individual adoption of these technologies are encouraged to search the Telemedicine Information Exchange (http://tie.telemed.org) for literature in these areas.

health scene has changed radically since the beginning of the 20th century. Technological innovations in the basic sciences and engineering have permeated almost every facet of modern life. The results of these inventions have been dramatically apparent in the delivery of health services. Although the art of medicine has a long history, the ability of a health care system to prevent and cure illnesses effectively is a relatively new phenomenon whose roots lie in the technological advances of the 20th century (Bronzino, Smith, & Wade, 1990).

Technology shapes medical care today. Health technologies have been defined as "the set of techniques, drugs, equipment, and procedures used by health care professionals in delivering medical care to individuals and the systems within which such care is delivered" (Office of Technology Assessment, 1976, p. 4). Health care scholars classify medical technologies into categories by purpose. Geisler and Heller (1996), for example, provide the following typology of health technologies:

1. *Medical devices*: equipment, instruments, machines, and other devices used for clinical diagnoses and care

2. *Drugs/pharmaceuticals*: compounds used in clinical care, provided via prescriptions or over the counter

3. *Disposables*: one-time usage materials and devices that are discarded after use and do not constitute equipment in the medical devices category

4. *Medical/surgical procedures and services*: medical and surgical knowledge involved in carrying out medical/surgical interventions

5. *Information technology*: informatics, automation, and computer usage classes of equipment, software, and techniques utilized in the clinical and the administrative areas of health care.

In attempting to understand the adoption and use of technology in health care, scholars have borrowed liberally from Rogers' (1995) classic treatise on the diffusion of innovations. The application of this theoretical framework in health is based on the notion that an innovation "refers to the emergence of novelty in the knowledge of medical science, practice, or organization … a change in the set of technologies available for the provision of health care "(Feeny, 1986, p. 5). Rogers' (1995) theory stipulates that the diffusion and rate of adoption for any innovation can be understood by studying its evolution through six steps from recognizing a need to understanding the consequences that result from the diffusion of a solution for this need. Rogers (1995) also stressed that diffusion is a communication phenomenon "in which participants create and share information with one another in order to reach a mutual understanding" (pp. 6–7). Some scholars have argued that Rogers' diffusion model is overly simplistic in its focus on a linear approach to understanding the diffusion of an innovation through a population. Whitten and Collins (1997), who view communication as participatory and local/conversational, argued against the appropriateness of linearly charting the curve of an innovation because the ability to isolate the specific innovation itself becomes problematic. They stated that the diffusion actually occurs at the level of conversations between people and often takes place in a radically decentralized manner with much reinvention along the way.

Application of diffusion theory in the study of health care technologies has been a fairly popular approach since the early 1970s. Feeny (1986), for example, developed a typology for health technology diffusion research that can be applied to any medical innovation:

- *Communication.* This area of research focuses on the flow of communication and information. Specific studies have analyzed channels of communication of innovations and differences in users' access to information regarding technologies.

- *Desperation-reaction.* Some researchers argue that diffusion and adoption are best predicted by understanding the initial efficacy expectations of a new treatment and the severity of the case.

- *Constrained optimization.* These studies focus on the role of the goals and constraints in the work of the health care provider who decides when and whether to adopt a new technology. Such research has documented that the relative advantage of adoption to a health provider/organization appears to be important in determining the rate of diffusion.

- *Administration.* With an emphasis on the organization, some researchers have focused on the effect of the administrative type and structure as well as individuals in leadership positions on the diffusion process.

- *Innovation cycle model.* This research approach focuses on the skills and learning of individuals or groups as determinants of the rate of adaptation and adoption of innovations.

Diffusion research has documented that health providers and patients have a love–hate relation with medical technologies. Technologies are credited with saving lives, improving health status, and improving the quality of health care. Simultaneously, technologies are vilified for playing a major role in the escalation of medical costs. Big-ticket items such as diagnostic imaging systems, new biotechnology products, or computer-based informatic systems attract both praise and blame for their contributions to the health sector.

Neumann and Weinstein (1991) outlined five facts about medical technologies that merit acknowledgment in any consideration of these technologies for senior citizens. First, new technologies do, on the average, improve the quality of medical care by improving health outcomes. One obvious example of this fact is the increase in the life span of older people. In 1900, 122,000 people were 85 years of age or older. By 1990, there were more than 3 million people 85 years of age or older (U.S. Bureau of the Census, 1996).

Second, many new technologies are ineffective or redundant and do not necessarily improve health outcomes. The challenge is the difficulty in determining the effectiveness of new technologies in a society that has no patience to wait for study and documentation. As Americans grow older, they want immediate relief for problems such as clogged arteries through instant medication or surgery. Only after much money has been spent in delivering these services do people begin to see the controversial research on the effectiveness of these innovations.

Third, new technologies do, on the average, add to health care costs. As technologies are rolled out into the homes of elderly citizens, the sheer amount of services they will receive is potentially increased. Some rural residents, for example, had no access to a specialist in rheumatology, so they never received these services. As access is increased, the level and quantity of service being provided may also be increased, resulting in a greater bottom line for health expenditures.

Fourth, the incentives and regulations built into the American health sector lead to the inappropriate diffusion of technologies through underdiffusion of effective and cost-effective technologies and overdiffusion of ineffective innovations. In 1999, the time this article was written, the Health Care Financing Administration (HCFA) reimbursement structure for telemedicine required that both a specialist and a referring physician/provider be present at a consultation for reimbursement to occur. Because the fee had to be split between the physicians, and because most physicians viewed it as a waste of precious time to "go with a patient to a specialist," these regulations were causing a disincentive for telemedicine diffusion. As a result, an elderly person who could have benefitted from visiting a geriatric psychiatrist via telemedicine might not have had the opportunity because her family practitioner might not have had the time to participate in a telemedicine visit that he would not have normally attended in traditional forms of health delivery.

Fifth and finally, the American public cannot get enough of new medical technologies. Society has come to expect technologies as part of medical care. One study looked at elderly patients' perceptions of telemedicine technologies that were placed in their homes so home visits could be performed via telemedicine. Although they liked the service, the elderly patients were nonchalant and even unimpressed with the technology (Whitten, Collins, & Mair, 1998).

Previously, the author provided a typology of health technologies in regard to their general purpose. This discussion seeks to focus specifically on medical technologies that have an impact on the delivery of medical services to the elderly. Stoeckle and Lorch (1997) described four types of medical technologies that play a role in assistance and self-care health functions: (a) *assistive* body equipment and devices that are used to improve the capacities of the physically or sensory disabled; (b) *testing* chemical kits and instruments used for monitoring the treatment of disease and for screening to detect disease or risks; (c) *treatment* medications as well as intravenous and enteral therapies used to provide acute and chronic disease management; and, (d) *telecommunications* that patients with providers and Internet information networks.

The remainder of this chapter focuses on telecommunication technologies, often referred to as telemedicine or telehealth, that are used to deliver medical services and education. It should be stressed, however, that any of the other categories listed by Stoeckle and Lorch (1997) and the typology provided by Geisler and Heller (1996) can be accessed through telecommunication technologies. For example, there are assistive body equipment devices, such as voice recognition systems, that assist the disabled in surfing Internet health sites. Also, there are Internet sites that enable consumers to purchase medical devices or pharmaceutical products. (chap. 14 provides additional discussion about the potential difficulties associated with older adults accessing health information on the World Wide Web.) With this caveat in mind, the discussion now turns to an overview of telemedicine.

Understanding telemedicine and its applications in health is important for several reasons. First, health care in America currently is undergoing powerful changes in both the structure and delivery of medical services. For example, increasing demands for managed care are creating new relationships among physicians, patients, health care providers, insurance companies, and government agencies, indicating the need for delivery systems that allow for full participation and access among the disadvantaged.

Second, recent advances in electronic telecommunications technologies present new opportunities for minimizing distances between and among caregivers, informational and educational services, and patients. The first survey of telemedicine programs found 12 active programs in the United States and Canada in 1993 (Allen & Scarbrough, 1996). Nearly all of these programs consisted of interactive video with room units or roll-about units. By 1998, hundreds of programs were doing patient–clinician interactions. This rapid growth implies that telemedical technologies are here to stay, and that there must be understanding about how they are being used to provide care for our senior citizens.

Third, Perednia and Allen (1995) predicted that most physicians in the United States would be directly or indirectly involved with some form of telemedicine by the year 2000. Indeed, almost all health providers have already felt the impact of information technologies and medical informatics. Therefore, practitioners of telemedicine must understand these similarities and differences as they gradually move from their traditional medical setting to one mediated by new technologies.

Taken together, these trends suggest the need for a thorough understanding of telemedicine, an important context in which health providers seek to adopt and adapt new information technologies to provide quality services to patients, particularly to those geographically isolated.

Telemedicine and Telecommunication Technologies

Telemedicine has been defined in a variety of ways ranging from very specific to more general. Park (1974) defined telemedicine very narrowly with a focus on the technology when he described it as the use of interactive or two-way television to provide health care. Higgins, Conrath, and Dunn (1984) also focused on technology in their definition of telemedicine as the use of telecommunications technology to assist in the delivery of health care (p. 285). Broader, less restrictive definitions, on the other hand, describe telemedicine in terms of the physician diagnosing or treating a patient who is at a different location (Willemain & Mark, 1971). Bennet, Rappaport, and Skinner (1978) expanded on this broader concept by using the term "telehealth," in which they included education, administration, and patient care.

Advances in the 1990s expanded the concept of telemedicine to include any technology using telecommunication technologies to deliver health services. This can include videoconferencing on personal computers, on larger room-based systems, or even through an individual's regular analog telephone lines. It also can include any applications of Internet telephony, which would include data, voice, and/or audio transmitted through the Internet. Until recent years, telemedicine traditionally meant the synchronous delivery of health services via interactive video or the asynchronous delivery via store-and-forward technologies.

TRADITIONAL TELEMEDICINE

The use of telecommunication technologies for medical diagnosis, care, and education has traditionally and commonly meant the use of interactive video for synchronous delivery of care. Interactive video services are fully synchronous. In this type of application, two or more parties are both physically present in front of the ITV equipment seeing and hearing each other. Of course, the quality of the interactions depends on the equipment and transmission speeds being used.

Telemedicine techniques, as defined previously, have been under development since the early 1960s. Wittson, Affleck, and Johnson (1961) were the first to use telemedicine for medical purposes in 1959 when they set up telepsychiatry consultations via microwave technology between the Nebraska Psychiatric Institute in Omaha and the state mental hospital 112 miles away (Jones & Colenda, 1997; Wittson et al., 1961). In the same year, Montreal, Quebec, was the site for pioneer teleradiology work being done by Jutra (1959). In the 1970s, there was a flurry of telemedicine activity as several major projects developed in North America and Australia, including the Space Technology Applied to Rural Papago Advanced Health Care (STARPAHC) project of the National Aeronautics and Space Administration (NASA) in southern Arizona; a project at Logan Airport in Boston, Massachusetts; and programs in northern Canada (Dunn et al., 1980). Although data are limited, early reviews and evaluations of these programs suggest that the equipment was reasonably effective at transmitting the information needed for most clinical uses, and that users for the most part were satisfied (Conrath, Puckingham, Dunn, & Swanson, 1975; Dongier, Tempier, Lalinec-Michaud, & Meunier, 1986; Fuchs, 1974; Murphy & Bird, 1974). Interestingly, with exception of one simple program at the Memorial University Hospital of Newfoundland, no telemedicine programs survived past 1986. When external sources of funding were withdrawn, the programs simply folded.

The decades of the 1960s, 1970s, and 1980s exhibited a series of telemedicine pilot and demonstration projects. However, the 1990s proved to be a period of rapid growth. In the early 1990s, new and fairly inexpensive digital technologies became commonly available that enabled video, audio, and other imaging information to be digitized and compressed. This facilitated the transmission of this information over land lines with relatively narrow bandwidths, instead of through the more expensive satellite technology or the relatively unavailable private cable or fiber optic lines. In 1990, there were four active telemedicine programs. By 1997, there were almost 90 such programs, and currently there are more than 200 documented telemedicine programs. It is worth noting that a few preliminary studies indicate that telemedicine is a viable alternative for health care treatment (Perednia & Allen, 1995). Indeed, research findings related to medical efficacy and satisfaction testify to the feasibility of this alternative. As Allen, Cox, and Thomas (1992) reported, "The telemedicine interaction was found to be a reasonable substitute for an on-site patient–physician encounter, in terms of patient–physician satisfaction and ability to transmit information and diagnosed" (p. 323).

Small-scale studies have been performed to assess patient satisfaction with telemedicine consultations and interpersonal issues in a variety of settings throughout the world (Allen, Hayes, Sadasivan, Williamson, & Wittman, 1995; Crichton et al., 1995; Dongier et al., 1986; Jerome, 1993; Pederson & Holand, 1995), indicating a generally positive response from patients for telemedicine. Patients have agreed that there are definite advantages to telemedicine, which include reduced waiting times, reduced costs to the health care system, impressions that examinations are more thorough, and excitement with the use of this new technology. Disadvantages that have been noted include nervousness about the use of the new technology, difficulty in talking to health care providers via the TV system, a tendency to be less candid in talking to the provider via this medium, and an experience of emotional distance between users and their providers (Allen et al., 1995; Crichton et al., 1995; Jerome, 1993; Dongier et al., 1986; Pedersen & Holand, 1995).

Traditional Telemedicine Projects for the Elderly

Since its introduction more than 40 years ago, telemedicine has been used for treating older adults. In the early 1970s, interactive cable was used between Mount Sinai Medical Center in New York City and a public housing high-rise for older adults, allowing tenants to ask questions and participate in discussions of nutrition, exercise, chronic diseases, and other health-related topics of interest to the elderly (JAMA, 1973). Traditional telemedicine programs designed specifically for older adults now encompass both synchronous interactive video (ITV) projects and asynchronous store-and-forward programs. With ITV projects, the health provider and patient see each other over a television or personal computer (PC) monitor and conduct a visit in real time.

With store-and-forward services, data is collected and reviewed at different times. The technology does not provide immediate feedback. Table 7.1 displays exemplar ITV and store-and-forward programs developed for older adults.

The bottom line is that there exists a variety of synchronous and asynchronous telemedicine projects designed for older adults. However, it is impossible to classify one method as superior to the others. Instead, the appropriate timing for a telehealth service is contingent on the urgency of the health condition as well as the available resources (including personnel, available time, and cash). The decision to provide telehealth services in real time rather than a store-and-forward format is based on the delicate balance between what is best for the patient and what resources are available for the providers.

E-health.   The latest trend in the field of telecommunication and health has been dubbed "e-health." In fact, some are suggesting that the term "e-health" will

TABLE 7.1

Exemplar Synchronous and Asynchronous Programs

Synchronous Interactive Video Programs	*Asynchronous Store-and-Forward Programs*
Fisk (1998) argued that personal monitoring devices should be considered as part of home health care in the United Kingdom. The author argued that because sensors are cheap this should facilitate adopting. He also argued that sensor technologies should improve health care because they will allow health care providers to intervene in a timely manner and help the elderly to maintain independence, thus facilitating acceptance among the elderly.	A telemedicine research project was established in New York state to address the challenges faced by vascular nurses. This project, run by Bassett Healthcare, a rural integrated health care network in central New York, went through two phases. The first phase involved gathering baseline data and conducting a needs assessment. In the second phase, the equipment was chosen and installed. A vascular nurse was invited to participate at this stage, who used the Picasso telephone technology (Lucent Technologies) for store-and-forward services. This led to the creation of the Vascular Nursing Teleconsultation Service. The researchers concluded that store-and-forward technologies are ideally suited for vascular nurses because they are already technology-savvy, and that this will only enhance patient care. (Lewis, McCann, Higdalgo, & Gorman 1997)
Because the elderly are susceptible to various skin diseases, telemedicine consultations offer a possibility for cost-effective diagnosis and treatment. In a project set in a nursing home, researchers conducted a small pilot study and concluded that most consultations were accurately diagnosed, and that for the skin conditions under study, most did not need the presence of a physician. However, various obstacles remained, primarily acceptance by the patients and reliance on standard telephone lines, which meant that the system was slow and incapable of handling large amounts of data at any given time. (Zelickson & Homan, 1997).	Older patients in Virginia who had undergone hip or knee replacement surgery and were taking blood-thinning agents to prevent the formation of blood clots were taught how to check their own coagulation rates at home using a home-monitoring device, then call in the readings to their health care professional. Nearly 100% of the participants in the study were satisfied with the process (Virtual Reality, 1998).

TABLE 7.1 (continued)

Exemplar Synchronous and Asynchronous Programs

Synchronous Interactive Video Programs	Asynchronous Store-and-Forward Programs
In France, elderly patients participated in teleconsultation with an orthopedic surgeon after surgery or traumatic injury. Patients had favorable attitudes and reported a high level of satisfaction with teleconsulting, whereas the surgeon generally felt a satisfactory level of confidence in decision making using teleconsulting (Couturier et al., 1998). In a separate project, elderly patients in France also participated in remote psychological consultation. Although the study showed some difficulty in assessment resulting from hearing impairment of the patient, it demonstrated the potential for remote consultation (Montani et al., 1996).	Research conducted at two ambulatory clinics in the state of New York compared otolaryngologic consultations using live two-way video and a store-and-forward service. The live two-way sessions were hampered by loss of color and inadequate lighting, whereas the store-and-forward service was greatly enhanced if an extended video clip was included. However, the authors argue that a store-and-forward service is more desirable because it enhances the level of care for routine consultations (nonemergency) by eliminating unnecessary travel for the patient, but there are numerous obstacles to overcome before this technology becomes commonplace. Among these are the ease of use, the incompatibility of hospital information systems, and technical training. The authors conclude that despite these obstacles, including a decrease in the lack of certainty, store-and-forward service have many potential benefits including helping rural and urban physicians to consult, improved access to emergency care physicians, and consultation between specialists and specialists. (Selafani, et al., 1999)
The National Institute on Alcohol Abuse and Alcoholism (NIAAA) identified the elderly as a special population at increased risk for alcoholism (Coogle, Osgood, Parham, Wood & Churcher, 1995), and found that practitioners, caregivers, and older adults themselves were unaware of the effects of alcohol on older people. The Virginia Geriatric Education Center effectively used videoconferencing to educate service providers, older adults, and family caregivers about geriatric alcoholism. This study demonstrated that practitioners and lay people can be educated simultaneously with a single outreach program, and showed the feasibility of including consumers of aging services among those targeted by geriatric education programs.	Kvedar et al. (1997) conducted a study that sought to compare agreement rates between live consultations and the use of store-and-forward technologies for older adult patients at the dermatology clinic at Massachusetts General Hospital in Boston.. The authors concluded that live two-way systems are very expensive, and that store-and-forward technology is a potential solution. They concluded that store-and-forward technologies were satisfactory, resulting in relatively high agreement between remote and office-based dermatologists. However, one of the drawbacks was image quality because most of the images were taken by untrained photographers (i.e. the physician with a digital camera).

TABLE 7.1 (continued)

Exemplar Synchronous and Asynchronous Programs.

Synchronous Interactive Video Programs	Asynchronous Store-and-Forward Programs
Central Kansas is the site for some groundbreaking home health services being delivered to the elderly. In this project, inexpensive based videoconferencing systems are put into the homes of patients. Nurses are then able to make home health visits to patients through the telemedicine system. Initial evaluation of this program has documented that it is an efficacious and satisfactory means for home health services (Whitten, Collins & Mair, 1997).	In Italy, the TeSAN company offers a call-center specifically tailored to the need of the elderly client. TeSAN offers three basic services for clients: (a) a personal emergency response system that allows homebound clients to call for the police, a physician, an ambulance, or other assistance, at the push of a button; (b) a careline service in which periodic proactive calls are made to subscribers to check on their status and assess their needs; and (3) a telemedicine/telemonitoring service for newly discharged hospital patients. Vital signs are taken by the patient or caregiver at the patient's home, and the information is relayed to the physician (Allen, Cristoferi, Campana, & Grimaldi, 1997).
In Japan, videophones using full-color motion pictures and sound have been used to assess at-home rehabilitation programs for the disabled elderly. Use of the videophone increased the communication abilities of the elderly, stimulated the patients' attention, and improved their comprehension and expression. Furthermore, the videophone improved patients' self-esteem and sense of belonging to society (Takano, Nakamura, & Akao, 1995).	Doctors at Purdue University tested a program in the early 1990s of in-home monitoring of patients suffering from congestive heart failure. Results of the study showed that patients could easily use the system, and physicians could effectively use the clinical data that was obtained (Patel & Babbs, 1992).
In Germany, elderly people have been connected via a broadband video communication system to a telecare center. The system, known as the TeleCommunity, is designed to allow elderly and mobility-impaired people to live independently while reducing the demand on social service resources (Erkert, 1997). Patients reported positive satisfaction with the system and considered it an "irreplaceable enrichment to their lives." The system gave many of the elderly clients a feeling of security and involvement. They reported that it was important for them to know daily contact was available to them if needed.	Currently in Wales, occupational therapists are using electronic sensors for automated assessment of elderly patients after illness or accident (Doughty & Costa, 1997). The sensors monitor activities of daily living such as bathing, cooking, dressing, eating, and mobility and use of the stairs in the home. This system allows monitoring of the conditions that may put the individual in an at-risk situation. Furthermore, by having the patient wear a sensor, the system is able to use simple algorithms to calculate the risk of the elderly person falling, allowing the health care staff to intervene before the fall takes place (Doughty & Cameron, 1998). This telemedicine system may allow the elderly person to return home sooner after illness or accident and permits interventions to be made before the individual is exposed to the risk of personal injury or illness.

replace the term telemedicine. E-health applications fall within the purview of a field dubbed interactive health communication, which is defined as the interaction of an individual (e.g., consumer, patient, caregiver, health provider) with or through an electronic device or communication technology to assess or transmit health information or to provide or receive guidance or support on a health-related issue (Robinson, Patrick, Eng, & Gustafson, 1998).

E-health, however, refers specifically to health-related services provided via the Internet. The e-commerce segment of the Internet health industry seems to be generating great excitement, particularly among Wall Street analysts and Internet health vendors. According to a recent study by Jupiter Communications, the online consumer health care market is expected to grow to $1.7 billion by 2003 (Industry Report: Internet Technologies in Healthcare,1999).

The health care industry, which now competes voraciously for every heath care dollar, is witnessing a convergence of care suppliers, payer sources, providers, and consumers. With this emerging interdependence comes the heightened need for ubiquitous access to information and services that help industry members confront some of the universal challenges, including pressures to reduce costs, intense competition, the complexities of managing information as people and companies integrate, and an ongoing need to gain and retain a competitive advantage through growth and diversification (Ernst & Young, 1997). As the lines that distinguish the various health sectors begin to blur, e-commerce emerges as a strategic resource for the purchase and sale of products and services to and between health providers. To understand e-health better as a portion of a burgeoning e-commerce environment, it is important to delineate the major types of health services currently found on the Internet, as in the following discussion. After a description of each type of Internet-based health service, several specific Web sites are provided as examples.

Medical Equipment and Supplies.   Health commodity items are perhaps the easiest to transition from traditional hard-copy catalogs to online publications. Traditionally, health providers must work through a company sales representative or distribution middleman to purchase medical equipment and supplies for their clinics or offices. The Web offers an opportunity for staff to shop on the Web for everything from hospital beds to bedpans and tongue depressors. This category refers to all items that are tangible and require physical transportation for delivery.

There are two potential separate sets of clients for medical equipment and supplies. The first are industry members: hospitals, doctors' offices, outpatient health centers, home health agencies, nursing homes, and medical laboratories. Any organization that provides direct or indirect patient services can purchase equipment and provide necessary to supply that care through these Web sites. The second set of clients for this category is composed of the care recipients. These would include the actual patient, a family member, or a nonprofessional caretaker of a patient. A host of medical equipment and supplies can be purchased directly by the actual patient: special-functioning beds, wheelchairs, canes, glucose monitors and strips, blood pressure monitors, bulk items such as wound care dressings and bandages, and so on. It appears that the preponderance of e-commerce sites currently being developed are directed at businesses rather than individuals, which is logical given current computer dissemination and use trends.

- CustomerLinks (*http://www.pssd.com*) bills itself as the first Internet-based health care solution designed for medical practices. CustomerLinks claims to be an information source that provides a multitude of information and services with 24-h access. Among the information and services available through this company is direct online ordering of medical supplies, equipment, and pharmaceuticals.

- Able Medical Aids (*http://www.ablemedical.com/default.asp*) provides medical equipment and supplies for clients to use in their homes, or assisted-living environments, and for professionals to use in the medical field. Thus, prospective clients range from individual consumers to health organizations. Products that can be purchased through the Internet from this company include medical equipment and supplies, bathroom safety equipment, mastectomy supplies, ostomy and incontinent supplies and medical equipment, and wheel chairs. Delivery is provided locally in the company's Florida location as well as nationally.

Clinical Services.   The Internet serves as an intriguing source for the actual acquisition of medical care and services. The purchasing of medical diagnoses, treatment recommendations, ongoing care management, or a simple second opinion from a licensed health provider falls into this category. These services theoretically could be purchased from a physician, nurse, nurse practitioner, physician assistant, psychologist, social worker, or speech/physical/occupational therapist.

Currently, the handful of forays into cybermedicine appear to be coming from a few enterprising physicians who have set up medical practices on the world wide web. In these practices, patients log onto the Internet, type in a description of their symptoms accompanied by a credit card number, and then are connected to a real doctor who offers a diagnosis as well as a prescription if warranted (Greene, 1997). This new service often is touted for its potential to solve the issue of access for people who are unable physically to see a physician in his or her office because of geographic, economic, or time constraints. However, the practice of medicine is a highly regulated business, and this method of service raises a number of perplexing questions ranging from licensure requirements to efficacy of care. Additionally, there are worries that Internet-based clinical service provision opens the door to increased probability of misrepresented health provider credentials, as well as patient attempts to fake illness in order to obtain prescription drugs.

- Cyberdocs (*www.cyberdocs.com*) is the Internet's dominant site for real-time online medical consultations. Founded in July 1999 by three emergency room doctors, Cyberdocs has a proven business model and the only Internet malpractice insurance policy ever written. A full 60% of Cyberdocs' demand for consultations comes from abroad. Substantial traffic to Cyberdocs' site has been generated without advertising or registering on a search engine.

Health Insurance.   Payers of health services obviously play a pivotal role in the entire health system. Currently, most health insurance companies are using the Web for informational purposes. However, some companies are using e-commerce in one of two ways. First, following the traditional independent agent structure of the insurance system, insurance companies are using the Web to enable agents to order policies electronically for their clients. Second, insurance companies are using the Internet to bypass independent agents and sell health insurance policies directly to the end consumer. Some sites provide consumers with electronic forms to speed up the process of obtaining health insurance. Other sites go a step further and actually support the online purchase of health insurance.

- The Provident health insurance agency (*http://www.theprovident.com*) with headquarters in Norristown, Pennsylvania, provides a Web site that offers general information about its major products. In addition, its site provides licensed agents with the tools necessary to complete a sales transaction. These tools include marketing updates and rate software.

- Champion Insurance Advantage Ltd. (*http://www.champion-ins.com*), located in Bel Air, Maryland, offers several types of health insurance policies online including travel medical insurance, study-abroad insurance for students who study in other countries, student health insurance for U.S. university and graduate students, and temporary health insurance for people between jobs or new U.S. university graduates. This Web site offers links to a summarization of each coverage type, a rate quote, and a location to which an application for insurance coverage can be submitted.

Medication.   Consumers use a wide range of prescription and over-the-counter medications for prevention and treatment purposes. Medications include anything traditionally provided in the health section of a supermarket or drug store or products supplied only through a physician's prescription. E-commerce sites offering medication products are emerging in three distinct forms: (a) online sales of over-the-counter medications delivered directly to the consumer; (2) online ordering of prescription medications that can be picked up by the consumer or delivered directly to the home or office, and; (c) direct marketing with online consultation service for a specific product available by prescription only.

- MoreOnline.com (*http://www.moreonline.com/index_medicine.html*) provides an online supermarket for consumers who wish to have over-the-counter products mailed directly to their homes. Located within the supermarket is an extensive health section from which consumers can purchase a wide range of over-the-counter medications including, but not limited to, analgesics, antacids, anti-diarrhea products, cold and allergy relievers, menstrual remedies, and vitamins. Clients simply load their shopping carts with these products and charge them to a credit card.

- In an attempt to improve customer service, Wal-Mart *(http://www. wal-mart.com/stores/pharm_mail.shtml)*, a national U.S. discount store chain, is offering an online service for the shipment of any prescription drug within 24 h of receipt. Customers simply e-mail their prescription (obtained from a physician) to Wal-Mart. Clients must provide the name of the medication, their doctor's name, and their doctor's phone number so Wal-Mart can verify the prescription.

- Medical Center.net *(http://www.medicalcenter.net/viagra.html)* is an Internet marketing and administrative agency owned by the Pill Pharmacy. It sells Viagra, Propecia, and Claritin to clients around the world through the Web. For example, clients wishing to obtain Viagra are able to obtain an online consultation for a prescription, which then is shipped directly to the client. A credit card is all that is needed to purchase 10 to 30 Viagra pills. The Web site is careful to include a clear disclaimer telling visitors that they should not take Viagra if they are currently taking any kind of nitrates or any of the drugs listed on a contraindicated list provided on the site.

- Incorporated in June 1998, Rx.com *(www.rx.com)* is an Internet-based retail pharmacy. Rx.com focuses exclusively on the sale of prescription pharmaceuticals; over-the-counter (OTC) pharmaceuticals; vitamins, minerals, supplements (VMS) and herbal products; and related health care products. Rx.com's online pharmacy allows customers via the Internet to receive comprehensive drug and health care information, and to correspond privately with its pharmacists.

Alternative Medicine.   Alternative medical solutions to traditional Western medical practices are increasing in popularity and acceptance. In 1996, the National Library of Medicine and the Medical Subject Headings Term Working Group, Office of Alternative Medicine, National Institutes of Health defined alternative medicine "as an unrelated group of nonorthodox therapeutic practices, often with explanatory systems that do not follow conventional biomedical explanations" (Alternative Medicine Home Page, February 1, 1999). Alternative medical therapies include folk medicine, herbal medicine, diet fads, homeopathy, faith healing, new age healing, chiropractics, acupuncture, naturopathy, massage, music therapy, and aroma therapy. The Internet provides a ready source for information about alternative sources of care for a wide range of chronic and acute health conditions such as acquired immunodeficiency syndrom (AIDS), arthritis, cancer, or back pain.

Traditionally the patient is the main seeker of information regarding these alternative sources of medical treatment. However, health providers also are displaying increasing interest in alternative sources of care. The November 1998 issue of the *Journal of the American Medical Association* (JAMA) highlights trends in alternative medicine use in the United States as well as the creation of a number of academic research centers for alternative medicine (e.g., University of Texas, University of Medicine and Dentistry of New Jersey, Bastyr University).

- Lehning Laboratories' homeopathic formulas have been available for almost 60 years through traditional channels. Currently, the company Enzymatic Therapy (*http://www.enzy.com/homeo/*) claims to be the exclusive distributor of these products that are available through health food retailers. Clients interested in purchasing these homeopathic remedies are able to locate the store nearest to them that offers these products through the company's Web site.

- Acupuncture.com (*http://www.acupuncture.com/marketpl.html*) offers consumers information about acupuncture and sells Chinese medical supplies for this alternative source of care. In addition, this site provides links to preferred provider organizations seeking licensed acupuncturists, malpractice providers for acupuncturists, and an acupuncture provider network.

Health Information. The Internet is growing rapidly as a popular source of medical information. Brown (1998) found that 43% of Internet users in 1997 accessed health or medical information. Currently, more than 10,000 health and medical sites exist on the World Wide Web (Hafner, July 9, 1998), most of which appear to emphasize access to health information rather than transactions. Insurance companies, medical suppliers, hospitals, clinics, laboratories, and even individual practitioners have created informational Web sites as an indirect marketing tool. Often, these sites simply provide information about the product, service, and provider, with the goal of persuading a client to use a facility or provider or to seek a specific product. In essence, the site serves the same functional role as a brochure, direct mailing piece, or newspaper advertisement. However, even in this seemingly enormous category of e-commerce sites, distinctions emerge in regard to purpose and goals.

- *Product/service information.* Many sites provide generic information about a specific product or service. These sites tend simply to introduce a product or service and explain its function. Many hospitals provide information about available services, operating hours, locations, and how one can gather further information (e.g., Mayo Clinic at *http://www.mayo.edu*, Cleveland Clinic at *http://www.ccf.org/*, or the Children's Hospital Boston at *http://www.childrenshospital.org/*).

- *Metasites that combine generic information with referral links.* A host of Web sites serve as gateways to specific providers or suppliers. Usually these sites provide generic information about an industry or service sector with a special link to a directory of providers. As with business directories in other industries, this combination of health or product information along with an online "yellow pages," introduces the benefits of lowered search costs to the health care sector. For example, the American Medical Association (*http://www.ama-assn.org/aps/amahg.htm*) home page provides information about the nonprofit group's mission and goals and various medical conditions and treatments. In addition, this organization provides an e-commerce service to its physician membership through an online doctor finder that directs Web site visitors to a specialist near them.

- *Education for health providers.* Many health organizations offer continuing education credits (mandated by U.S. states for ongoing licensure) for physicians, nurses, and allied health providers as a marketing tool. For example, hospitals hope that physicians will refer patients to them for tests and procedures; pharmaceutical companies hope that physicians will prescribe their medications; and nonprofit agencies hope that nurses or physicians will retain their membership. In addition, many academic institutions are viewing online continuing education as a potential revenue stream in its own right. The Internet offers interested parties a way to make continuing education credits available on a continuous basis without necessitating the expense of physically bringing experts to the students. For example, Stanford University (*http://www-radiology.stanford.edu*) provides continuing medical education for radiologists through their Web site for a fee. Radiologists select the topic of interest, register and pay the stipulated fee to participate, and obtain credit.

In summary, health-related services delivered via the Internet display enormous potential as the predominant form of telemedicine for the future. Yet, it is important to ensure that the biggest population segment in terms of health care dollars spent is not ignored as policy is created and e-health applications are designed. Since its introduction into mainstream America, the Internet has been touted as one of the fastest diffusing technologies to date (Atkin, Jeffres, & Neuendorf, 1998). According to data gathered during May 1999, the Internet has more than 61 million users spending an average of 7 hours online monthly (Nielsen//NetRating, 1999). Also, during this time, users averaged 16 sessions and visited 21 sites per session. Each of these sessions averaged 26 min and occurred evenly throughout each day of the month. Whereas previous research has indicated a large gender gap, current numbers suggest that there is a virtual split between female (49%) and male (51%) users (Nielsen NetRatings, 1999).

Americans are going online in increasing numbers for health-related information. In 1998, 60 million online searches involved health and medical information

(Harris & Associates, 1999). The National Library of Medicine, which opened its Medline database to the public 2 years ago, reported that searches increased from 7 million in 1997 to 120 million in 1998, with one third of these searches coming from consumers. Whereas these statistics are interesting, an examination of a specific meta health site such as Centers for Disease Control (CDC) home page is even more compelling. Visitors to the CDC Web site spent an average of 6 min 40 secs at the site while visiting an average of five pages. Of these visits, hepatitis was the number one disease-specific site visited (7.3%) followed closely by diabetes (5.4%). According to the Nielsen NetRatings statistics for May 1999, 314,000 people older than 55 years visited the CDC Web site, comprising 11% of their total visitors.

Missing from Internet health research to date have been analyses on use preferences for different audiences. It is known that older adults spend much of their time with mass media, using the media for a variety of reasons including these: to keep active, to gain information about the world in general, to pursue agenda-setting purposes, to experience companionship, to maintain interest in the world, and to pass the time (Barton & Schreiber, 1978). Yet, whatever the purpose of the mass media interaction, older persons are particularly drawn to news and information content, regardless of the mass media type (Tamir, 1979). This makes clear the potentially attractive role the Internet will play as a source of health information and services for older adults.

Because researchers have documented some cognitive and communicative differences between young and old adults, it is important to understand any differences in Internet usage and preferences by the elderly. For example, research in cognition as it relates to language and communication skills (Qualls, Harris, & Rogers, chap. 4, this volume) has documented changes in the speed of processing skills (Fisk & Fisher, 1994; Salthouse, 1996; Welford, 1985), auditory perceptual skills (Kline & Scialfa, 1997), and attentional skills (Plude & Hoyer, 1985).

These cognitive variations have some impact on older citizens' ability to understand and use spoken language, and on their reading and writing skills. The elderly show little change in sustained attention (vigilance) that requires paying attention to one task only. However, they do appear to have greater difficulty in allocating attention to the target task (Plude & Doussard-Roosevelt, 1989). Rogers (2000) pointed out, however, that the impact of age on attention skills varies by the specific type of attention. According to her findings, aspects of attention that decline for older adults include selective attention, divided attention, and the transition from attention-demanding to automatic processes.

Experimental evidence suggests that several types of memory are compromised in the normal elderly, but that these losses rarely do more than annoy or inconvenience the individual in everyday life (Maxim, 1994). However, the ability to manipulate information in the working memory may decline with age (Maxim, 1994). Research on diminishing problem-solving skills in the aging population has reported contradictory results. However, longitudinal studies and those that use educationally well-matched groups suggest that age differences may be quite small, and that miscellaneous variables such as education, task familiarity, and health status may have larger actual impacts (Charness, 1985). Gardner-Bonneau & Gosbee (1997) pointed out that memory is significantly involved in health behaviors related to the use of new technologies and medical devices. Smith (chap. 3, this volume) delineated important conclusions regarding memory aging and health-care behaviors such as the ability to reduce differences in memory with enhanced environmental support techniques.

Recently, there has been an increase in publications documenting the growing use of the Internet by senior citizens. One line of research in seniors' use of the Internet focuses on social impacts. For example, Cody, Dunn, Hoppin, and Wendt (1999) conducted a study of almost 300 older adults and found that those who actually learned to surf the Internet had more positive attitudes toward aging, higher levels of perceived social support, and higher levels of connectivity. They also found that the elderly surfers spent more time online when computer efficacy was high, computer anxiety was low, and attitudes toward aging were positive.

The more popular trend in elderly Internet publications seeks to document actual use trends. One study by Excite and Third Age Media found that 83% of online seniors log in daily, a number just shy of the 85% for 24 to 29-year-olds, the most active age group (Whelan, 1998). A host of computer and Internet classes targeting the elderly cannot accommodate all the interested students, and retirement communities are beginning to provide computers and Internet access as a marketing strategy. SeniorNet, a San Francisco-based, nonprofit group, already has taught more than 100,000 seniors to use computers and the Internet, and now has 141 learning centers around the United States (Whelan, 1998). SeniorNet estimates that within the next few years, 100 million U.S. citizens older than age 50 years will be online.

Although there are a host of publications about seniors' use of the Internet, very few appear to be grounded in empirical data. For example, one article claims that seniors prefer to be lured by peer-focused sites that feature chat rooms, lifestyle information, and targeted services (Capturing Seniors Online, 1997). However, there is no evidence that experimental data were used to draw these generalizations. Yet Internet marketers are running with these claims and having apparent success. A number of health care providers and/or marketers are diving into Web site development or sponsorship. Kaiser, for example, is sponsoring Ask the Pharmacists, a question and answer section that uses its pharmacists to answer prescription drug questions from SeniorNet subscribers.

THE FUTURE OF TELECOMMUNICATION TECHNOLOGIES FOR THE ELDERLY

The first wave of telemedicine, both interactive video and store-and-forward applications, have pointed to a variety of lessons. It is important to note that actual research for these projects since the early 1960s has been extremely weak and rarely grounded in empirical data. For example, Mair and Whitten (2000) conducted a systematic review of telemedicine satisfaction research and found that most of the research performed in this topic area lacks any consistent methodologic approach. As a result, it is difficult to conclude whether patients and providers are satisfied with telemedicine. However, they postulated that there is a bigger problem in the satisfaction literature than the quality of the research to date. Instead, the bigger question is whether the results from a specific telemedicine project actually can be generalized across all telemedicine contexts. Their report argued that research should focus on specific questions of interest rather than continue the tradition of generic satisfaction research to gain specific knowledge that will inform the field of telemedicine as a whole.

In a similar project that sought to analyze cost-effectiveness research in the field of telemedicine, Whitten et al. (1999) performed a search of six well-known databases with a variety of relevant key words related to telemedicine and cost effectiveness. After discarding non-English publications, books, and duplicate publications resulting from the same study they were left with 551 articles for analysis. Their second step was to separate the articles into two groups: those with and those without quantitative cost data. Only 38 articles contained any type of real data. Because many of these 38 studies proved to be inadequately designed or conducted, they were unable to perform a traditional meta-analysis. Furthermore, there were a number of disturbing features common to these studies, including the commonly omitted number of consultations or patients, the almost nonexistent longitudinal data collection, and the lack of uniformity in cost analysis. They concluded that it is premature for any statements to be made, either positively or negatively, regarding the cost effectiveness of telemedicine in general.

Although research in the field of traditional telemedicine has been little more than descriptive exposes of specific telemedicine projects, researchers are poised to prevent this from occurring in the coming generation of telecommunication technologies for health, namely e-health. E-health is gaining a foothold in the heath care industry despite the significant impediments that exist. Many of the emerging e-health applications now available on the Web point out the need for new research. In particular, three broad areas of evaluation require attention: the interaction between e-health and the structure of the health system, the role that e-health plays in the dynamics of health care delivery, and the impacts of e-health on older individuals and society. This chapter concludes with representative questions for each of these areas of evaluation.

System Structure

- How does electronic commerce operate within the current health care system? How do current providers, payers, and clients incorporate e-health into the acquisition of products and services? Will e-health serve to reinforce existing supplier–provider–payer–patient relationships or loosen them? How will Medicare respond to additional services available to older adults via the Web?

- Will e-health provide an impetus for health system changes and reform? Will e-health encourage the delivery of services directly from providers to clients, or will new health intermediaries emerge on the Web?

Dynamics of Delivery

- Do e-health sites and services need to be tailored for older adults? Will older adults require different system functionality and design than younger adults?

- Will e-health transactions encourage new consumer priorities and demands, such as instant access to medical consultations, or more convenient purchasing of health products? Will e-health encourage patients to seek distant medical care at the expense of local providers, or will it enhance service provision within a local context?

- Are there successful templates for e-health sites? What design elements enhance credibility and ease of use? How should the success of e-health sites be measured: by revenue, patient access, or health outcomes?

- Can e-health applications provide adequate information for effective consumer search and evaluation in health care? Will it transform health services from credence or experience goods into search goods?

- How are issues of security and confidentiality resolved? Will federal policy and regulations dictate these issues or will they be market driven? Will these solutions be based on technical or ethical solutions?

- Will medical practice regulations such as licensure and liability be impacted by the Internet? Will consumers lead the push to alter traditional state-restricted guidelines.

- Who will monitor e-health sites to maintain the integrity and validity of these businesses? How will we be sure that qualified vendors and clinicians are providing products and services? Will health care electronic markets be biased or unbiased?

Impacts and Outcomes

- Will e-commerce alter the main cost drivers in health care delivery? What costs will increase? What costs will decrease?

- Are clinical services delivered over the World Wide Web efficacious? Will health outcomes be impacted as clinical care and services are made more available via the Internet?

- Will there be any harmful effects to society as access to such health products as drugs is facilitated? What are the societal costs and benefits associated with reduced in-person health contact?

- What will be the overall impact on an economy's gross domestic product? Will e-health increase or decrease health-related spending? Will e-health shift the way dollars are spent within the health sector?

The use of telecommunication technologies to deliver services to older adults is not a new concept. In reality, the actual technologies have been around for a number of years. However, since the early 1990s, we have witnessed a growing number of health providers looking to telemedicine as a solution to make care more accessible. Unfortunately, the research during the past decade has been woefully inadequate to inform us regarding the clinical efficacy, cost benefits, or delivery issues related to telehealth. With e-health, or the full-scale delivery of many health services on the Internet, it becomes all the more important that researchers validate this delivery method because so many older adults will have easy access to the Internet. However, it must not be forgotten that any health service must operate within the constraints of a country's health infrastructure. If providers cannot get reimbursed for e-health services, or if they are uncomfortable with the liability risks, they will not provide care through these technologies even though this may be the only way older adults can access some of these services. Researchers in this field have their work cut out for them because they must balance traditional evaluation questions within a complex and dynamic health infrastructure.

REFERENCES

Allen, A., Cox, R., & Thomas, C. (1992). Telemedicine in Kansas. *Journal of the Kansas Medical Society, 93*, 323–325.

Allen, A., Cristoferi, A., Campana, S., & Grimaldi, A. (1997). TeSAN personal emergency response system and teleservices. *Telemedicine Today, 5*(6), 25, 33.

Allen, A., Hayes, J., Sadasivan, R., Williamson, S. K., & Wittman, C. (1995) A pilot study of the physician acceptance of tele-oncology. *Journal of Telemedicine and Telecare 1*(1), 34–37.

Allen, A., & Scarbrough, M. L. (1996). Third annual program review. *Telemedicine Today, 4*(4),10–13.

Alternative Medicine Home Page. (1999). Pittsburgh, PA: University of Pittsburgh. Available: *http://www.pitt.edu~cbw/altm.html*. Accessed: February 1, 1999.

Atkin, D., Jeffres, L., & Neuendorf, K. (1998). Understanding Internet adoption as telecommunications behavior. *Journal of Broadcasting & Electronic Media, 42*, 475–490.

Barton, B., & Schreiber, E. (1978). Media and aging: A critical review of an expanding field of communication research. *Central States Speech Journal 29*, 173–186.

Bennet, A. M., Rappaport, W. H., & Skinner, E. L. (1978). Telehealth Handbook. (Publication No. PHS 79–3210). Washington, DC: U.S. Department of Health, Education and Welfare.

Blazer, D. G., Landerman, L. L., Fillenbaum, G., & Horner, R. (1995, October). Health services access and use among older adults in North Carolina: Urban vs. rural residents. *American Journal of Public Health, 85*(10), 1384–1390.

Bronzino, J. D., Smith, V. H., & Wade, M. L. (1990). *Medical technology and society: An interdisciplinary perspective.* Cambridge, MA: MIT Press.

Brown, M. (1998). The HealthMed retrievers: *Profiles of consumers using online health and medical information.* New York: Cyber Dialogue.

Capturing seniors online: They're wired and ready for health information. (1997). *Healthcare PR and Marketing News, 6*(23), 1.

Charness, N. (Ed.) (1985). Aging and problem solving performance. In W. Charness (Ed.), *Aging and human performance* (pp. 225–259). Chichester, UK: Wiley.

Cody, M., Dunn, D., Hoppin, S., Wendt, P. (1999, April). Silver Surfers: Training and evaluating Internet use among older adult learners. *Communication Education*.

Conrath, D. W., Puckingham, P., Dunn, E. V., & Swanson, J. N. (1975). An experimental evaluation of alternative communication systems as used for medical diagnosis. *Behavioral Science, 20*, 296–305.

Coogle, C. L., Osgood, N. J., Parham, I. A., Wood, H. E., & Churcher, C. S. (1995). The effectiveness of videoconferencing in geriatric alcoholism education. *Gerontology and Geriatrics Education, 16*(2), 73–83.

Couturier, P., Tyrrell, J., Tonetti, J., Rhul, C., Woodward, V., & Franco, A. (1998). Feasibility of orthopaedic teleconsulting in a geriatric rehabilitation service. *Journal of Telemedicine and Telecare, 4*(1), 85–87.

Crichton, C., Macdonald, S., Potts, S., Syme, A., Toms, J., McKinlay, J., Leslie, D., & Jones, D. H. (1995) Teledermatology in Scotland. *Journal of Telemedicine and Telecare 1*(3), 185. [letter]

Dongier, M., Tempier, R., Lalinec-Michaud, M., & Meunier, D. (1986). Telepsychiatry: Psychiatry consultation through two-way television: A controlled study. *Canadian Journal of Psychiatry, 31*, 32–34.

Doughty, K., & Cameron, K. (1998). Continuous assessment of the risk of falling using telecare. *Journal of Telemedicine and Telecare, 4*(1), 88–90.

Doughty, K., & Costa, J. (1997). Continuous automated telecare assessment of the elderly. *Journal of Telemedicine and Telecare, 3*(1), 23–25.

Dunn, E., Conrath, D., Acton, H., Higgins, C., Math, M., & Bain, H. (1980). Telemedicine links patients in Sioux Lookout with doctors in Toronto. *Canadian Medical Association Journal, 22*, 484–487.

Erkert, T. (1997). High-quality links for home-based support for the elderly. *Journal of Telemedicine and Telecare, 3*(1), 26–27.

Ernst & Young L L P. (1997). *The role of the Internet in health care: Current state.* New York: Authors. [Brochure]

Feeny, D. (1986). Introduction: Health care technology. In D. Feeny, G. Guyatt, & P. Tugwell (Eds.), *Health care technology: Effectiveness, efficiency and public policy* (pp. 5–24). Montreal, Quebec: The Institute for Research on Public Policy.

Fisk, M. J. (1998). Telecare at home: Factors influencing technology choices and user acceptance, Journal of Telemedicine and Telecare, 4, 80–83.

Fisk, A. D., & Fisher, D. L. (1994). Brinley plots and theories of aging: the explicit, muddled, and implicit debates. *Journals of Gerontology: Psychological Sciences, 49*, 81–89.

Fuchs, M. (1974). Provider attitudes toward STARPAHC: A telemedicine project on the Papago reservation. *Medical Care, 17*, 59–68.

Gardner-Bonneau, D., & Gosbee, J. (1997). Health care and rehabilitation. In A. D. Fisk & W. A. Rogers (Eds.), *Handbook of human factors and the older adult* (pp. 231–256). New York: Academic Press.

Geisler, E., & Heller, O. (Eds.). (1996). *Managing technology in healthcare.* Boston: Kluwer Academic Publishers.

Greene, J. (1997). Sign on and say "ah-h-h-h-h." *Hospitals and Health Networks, 71*(8), 45–46.

Hafner, K. (1998, July 9). Can the Internet cure the common cold? *New York Times*, pp. D1, D7.

Harris and Associates, (1999). *Sixty million seek health info online in the U.S.* [Online] Available: http://www.nau.ie/surveys. Accessed: Feb. 2000.

Health Care Financing Administration. (1999). *Highlights of the national expenditure projections, 1997–2007*, Available: http://www.hcfa.gov/stats/nhe-proj/hilites.mtm. Accessed: February 1, 1999.

Higgins, C. A., Conrath, D. W., & Dunn, E. V. (1984). Provider acceptance of telemedicine systems in remote areas of Ontario. *Journal of Family Practice, 18*(2), 285–289.

Industry Report: Internet Technologies in Healthcare. (1999). *Healthcare Informatics,* 3–28, McGraw-Hill.

JAMA. (1973). Cable TV links hospital, apartments. *JAMA, 226*(12), 1410.

Jerome, L. (1993, January). Assessment by telemedicine. *Hospital and Community Psychiatry 44*(1), 81. [letter]

Jones, B. N., & Colenda, C. C. (1997, June). Telemedicine and geriatric psychiatry. *Psychiatric Services, 48*(6), 783–785.

Jutra, A. (1959). Teleroentgen diagnosis by means of videotape recording. *AJR American Journal of Roentgenology, 82,* 1099–1102.

Kline, D. W., & Scialfa, C. T. (1997). Sensory and perceptual functioning: Basic research and human factors implications. In A. D. Fisk & W. A. Rogers (Eds.), *Handbook of human factors and the older adult* (pp. 27–54). New York: Academic Press.

Kvedar, J. C., Edwards, R. A., Menn, E. R., Mofid, M., Gonzalez, E., Dover, J., & Parrish, J. A. (1997). "The substitution of digital images for dermatological physical examinations." *Archives of Dermatology, 133,* 161–167.

Lewis, P., McCann, R., Hidalgo, P., & Gorman, M. (1997). Use of store and forward technology for vascular nursing teleconsultation service. *Journal of Vascular Nursing, 25*(4), 116–123.

Mair, F., & Whitten, P. (2000). Systematic review of studies of patient satisfaction with telemedicine. *British Medical Journal, 320,* 1517–1520.

Maxim, A. (1994). *Language of the elderly: A clinical perspective.* San Diego, CA: Wherr Publishers.

Montani, C., Billaud, N., Couturier, P., Fluchaire, I., Lemaire, R., Malterre, C., Lauvernay, N., Piquard, J. F., Frossard, M., & Franco, A. (1996). Telepsychometry: A remote psychometry consultation in clinical geronotology: Preliminary study. *Telemedicine Journal, 2*(2), 145–150.

Murphy, R. L. H., & Bird, K. T. (1974). Telediagnosis: A new community health resource. *American Journal of Public Health, 64,* 113–119.

Neumann P. J., & Weinstein, M. C. (1991). The diffusion of new technology: Costs and benefits to health care. In A. C. Gelijns & E. A. Halm (Eds.), *Medical innovation at the crossroads: The changing economics of medical technology* (Vol. 2, pp. 21–34). Washington, DC: National Academy Press.

Nielsen//NetRatings. (1999). *Nielsen NetRating.* [Online] Available: http://www.nielsen-netratings.com. Accessed: February 2, 2001.

Office of Technology Assessment (OTA). (August, 1976). *Development of medical technology,* Washington, DC: U.S. Government Printing Office, OTA-H-34.

Park, B. (1974). *An introduction to telemedicine: Interactive television for delivery of health services.* New York: Alternate Media Center, New York University.

Patel, U. H., & Babbs, C. F. (1992). A computer-based, automated, telephonic system to monitor patient progress in the home setting. *Journal of Medical Systems, 26*(3/4), 101–112.

Pedersen S, & Holand, U. (1995) Tele-endoscopic otorhinolaryngological examination: preliminary study of patient satisfaction. *Telemedicine Journal 1*(1): 47–52.

Perednia, D. A., & Allen, A. A. (1995). Telemedicine technology and clinical applications. *JAMA, 273*(6), 483–488.

Plude, D., & Doussard-Roosevelt, J. (1989). Aging selective attention and feature integration. *Psychology and Aging, 4*, 98–105.

Plude D., & Hoyer, W. (1985). Attention and performance: Identity and localizing deficits. In W. Charness (Ed.), *Aging and human performance* (pp. xx–xx). Chichester: Wiley.

Robinson, T. N., Patrick, K., Eng, T. R., Gustafson, D. (1998). An evidence-based approach to interactive health communication: A challenge to medicine in the information age. *Journal of the American Medical Association, 280*, 1264–1269.

Rogers, E. M. (1995). *Diffusion of innovations* (4th Ed.). New York: The Free Press.

Rogers, W. A. (2000). Attention and aging. In D. C. Park & N. Schwarz (Eds.), *Cognitive aging: A primer* (pp. 57–73). Philadelphia: Psychology Press.

Salthouse, T. (1996). The processing-speed theory of adult age differences in cognition. *Psychological Review, 103*, 403–428.

Selafani, A. P., Heneghan, C., Ginsburg, J., Sabini, P., Stern, J., & Dolitsky, J. N.. (1999). Teleconsultation in otolaryngology: Live versus store and forward consultations, *Otolaryngology: Head and Neck Surgery, 120*(1), 62–72.

Stoeckle, J. D., & Lorch, S. (1997). Why go see the doctor? Care goes from office to home as technology divorces function from geography. *International Journal of Technology Assessment in Health Care, 13*(4), 537–546.

Takano, T., Nakamura, K., & Akao, C. (1995). Assessment of the value of videophones in home health care. *Telecommunications Policy, 19*, 241–248.

Tamir, L. (1979). *Communication and the aging process: Interaction throughout the life cycle.* New York: Pergamon Press.

The Internet Index [Online] Available: http//www.openmarket.com/inindex/99–05.htm. Accessed: Feb. 2000. Virtual Reality. (1998). Coagulation rates checked at home. *Telemedicine and Virtual Reality, 3*(5), 52.

Welford, A. (1985). Changes of performance with age: An overview. In W. Charness (Ed.), *Aging and human performance* (pp. 333–369). Chichester: Wiley.

Whelan, C. (1998, July 27). A computer for grandma. *Electronic News, 44, 222a,* 44–45. New York.

Whitten, P., & Collins, B. (1997). The diffusion of telemedicine: Communicating an innovation. *Science Communication, 19*(1), 21–40.

Whitten, P., Collins, B., & Mair, F. (1998). Nurse and patient reactions to a developmental telehome health system. *Journal of Telemedicine and Telecare, 4*(2)1.1–1.9.

Whitten, P., Kingsley, C., & Grigsby, J. (1999). Results of a meta-analysis of cost-benefit research: Is this a question worth asking? Paper presented at TeleMed 99, London, December.

Willemain, T. R., & Mark, R. G. (1971). Models of health care systems. *Biomedical Science Instrument, 8*, 9–17.

Wittson, C. L., Affleck, D. C., & Johnson, V. (1961). Two-way television group therapy. *Mental Hospital, 12*, 2–23.

U.S. Bureau of the Census. (1996). *Current population reports, special studies, P23–190, 65 + in the Unites States.* Washington, DC: U.S. Government Printing Office.

Zelickson, B., & Homan, L. (1997). Teledermatology in the nursing home, *Archives of Dermatolology, 133*, 171–174.

8

Overview and Intervention Approaches to Family Caregiving: Decomposing Complex Psychosocial Interventions

Richard Schulz
University of Pittsburgh

Sara Czaja
University of Miami

Steven Belle
University of Pittsburgh

The provision of assistance and support by one family member or friend to another is a pervasive aspect of everyday human interactions. Providing help to a family member with chronic illness or disability is not very different from the tasks and activities that characterize interactions among families and close friends without illness or disability. Therefore, when a wife provides care to her husband with Alzheimer's disease by preparing his meals, it may be an activity she normally would do for an unimpaired husband. However, if a wife also assists her cognitively impaired husband with bathing and dressing, few would question whether caregiving is taking place. The difference is that providing assistance with bathing and dressing or assisting with complex medical routines clearly represents "extraordinary" care and exceeds the bounds of what is "normative" or "usual." Similarly, parents caring for a child with a chronic illness may need to assist with daily medical routines (e.g., insulin injections or chest physical therapy) that are time-consuming, difficult, and added to normal parenting responsibilities. Caregiving involves a significant expenditure of time and energy, often for months or years, requiring the performance of tasks that may be physically demanding and unpleasant, frequently disrupting other family and social roles of the caregiver.

Although caregivers may carry out tasks similar to those carried out by paid health professionals, they perform these services for no compensation and do so either voluntarily or because they feel there are no other alternatives. Because the physical and mental health consequences of taking on this role are sometimes severe, and because caregivers represent an invaluable resource to the well-being of society, research on caregiving has become a high priority among scholars in many disciplines and among policymakers.

PREVALENCE OF CAREGIVING

Although the definition and boundaries of what is meant by the term "caregiving" often vary depending on the purpose for the definition, there is strong consensus that, regardless how caregiving is defined, its prevalence is high. A broadly inclusive approach might argue that a caregiver is needed for every person with health-related mobility and self-care limitations that make it difficult for the person to take care of personal needs such as dressing, bathing, and moving around the home. Current estimates indicate that 4% of the noninstitutionalized U.S. population younger than 55 years meet these criteria. Beyond the age of 55 years, the proportion of persons with mobility and/or self-care limitations increases dramatically, fully half of the population falls into this category after the age of 85 years. If it is assumed that these individuals minimally require one caregiver, these estimates yield more than 15 million caregivers in the United States. Indeed, these estimates are somewhat lower than the results from a recent national survey of caregivers reporting that 22.4 million households met broad criteria for the presence of a caregiver in the past 12 months (National Alliance for Caregiving, and the American Association of Retired Persons, 1997).

Caregiving is not just a late-life phenomenon involving the care of disabled older persons. It is estimated that 10% to 14% of children and adolescents (7.5 million) in the United States have some type of chronic illness or disability. Of these individuals, approximately 20% to 25% (1.5 million) have serious health conditions that impair daily functioning and thus require a caregiver. Additionally, 4.1 million individuals between the ages of 21 and 64 years require personal assistance in activities of daily living (ADLs) or instrumental activities of daily living (IADLs).

WHO PROVIDES CARE, AND WHAT TYPE
OF CARE IS PROVIDED?

Caregivers assisted elderly individuals generally are differentiated by age and relationship to the care recipient. One distinct subgroup of caregivers consists of adult children, usually daughters or daughters-in-law, in their 50s and 60s. The second group of caregivers, which involves spouses of care recipients, generally is older and includes a higher proportion of male caregivers than is found among adult child caregivers.

The roles and functions of family caregivers vary by type and stage of illness, and include both direct and indirect activities. Direct activities can include provision of personal care assistance such as administering help with bathing, grooming, dressing, or toileting; health care assistance such as catheter care, giving injections, or monitoring medications; and checking and monitoring tasks such as continuous or periodic supervision and telephone monitoring. Indirect tasks include care management such as locating services, coordinating service use, monitoring services or advocacy; and households tasks such as cooking, cleaning, shopping, money management, and transportation of the family member to medical appointments or day care programs (Biegel & Schulz, 1999). The intensity of these caregiving activities varies widely, with some caregivers having only limited types of involvement for a few hours per week and other caregivers providing more than 40 h a week of care and on call 24 h per day.

THE DEVELOPMENT AND EVOLUTION OF CAREGIVING RESEARCH

Thousands of studies on caregiving have been conducted by researchers from all of the social science and many of the health science disciplines. The research has progressed from simple, often cross-sectional, descriptive accounts of caregiving to complex hypothesis testing, longitudinal and quasi-experimental studies, and more recently, to intervention studies aimed at affecting caregiving outcomes.

Importantly, each patient population poses distinct caregiving challenges. With few exceptions, most caregiving studies focus on patients within a particular illness or disability category. This approach acknowledges the importance of disease type in shaping patient and caregiver outcomes, but it also makes comparative analysis across illness conditions difficult. If the level of disability is held constant and the disease context within which it occurs is varied, how does that affect caregiver outcomes? It might be hypothesized, for example, that providing personal assistance to a stroke patient with a stable prognosis is likely to be very different from giving the same amount of assistance to an Alzheimer's patient with an unpredictable and negative prognosis. Questions such as these are best addressed by simultaneously studying different patient populations using similar methods and measurement tools. Because macro-level policy is unlikely to be made at the level of specific diseases, it is all the more important that studies separate the disease-specific from general aspects of caregiving.

Like the patients, caregivers vary in age, gender, and relationship to the care recipient. The literature includes adult parents of young children, middle-age daughters caring for their parents or adult children, and older spouses caring for a husband or wife. Because gender, age, and relationship are often confounded, one of the challenges facing researchers is to disentangle the role that each of these factors might play in the type of help provided and the caregiving outcomes experienced. This complexity is compounded if add ethnic identity is added as another factor to this mix. For example, national data indicate that African American and Hispanic caregivers are more likely to provide more challenging personal care (e.g., high levels of assistance with activities of daily living [ADLs]) and to experience greater financial hardship than Asians or Whites (National Alliance for Caregiving, 1997).

THEORETICAL MODELS OF CAREGIVING IMPACT

Clinical observations and numerous empirical studies have highlighted the fact that caregivers react with marked individual differences to seemingly similar circumstances. Some family caregivers react with guilt, depression, and poor health, whereas other families provide care with either no ill effects or even positive consequences. Therefore, theorists have developed models to explain individual differences in caregiving. Much of the literature on caregiving can be characterized as an attempt to link some antecedent variables to outcomes assessing the well-being of individuals who provide support to ill relatives. Common independent variables in this concept are the functional or behavioral status of the patient, with a dependent variable assessing the psychosocial status or physical and mental health of the caregiver such as morale, life satisfaction, depression, or perceived strain or burden. A large number of individual and situational conditioning variables characteristic of all stress-coping models may moderate or mediate the relation between stressors and caregiver well-being. Examples include age, gender, socioeconomic status, type and quality of relationship between caregiver and patient, social support, and personality attributes of the caregiver, such as self-esteem and locus of control. The need for conditioning or intervening variables is justified by data demonstrating only moderate relations between patient impairment (e.g., ability to perform ADL tasks) and caregiver outcomes, such as mental health (Schulz, O'Brien, Bookwala, & Fleissner, 1995).

A number of theorists have proposed stress process models that share common features (Cohler, Groves, Borden, & Lazarus, 1989; Haley, Levine, Brown, & Bartolucci, 1987; Montgomery, Stull, & Borgatta, 1985; Schulz, Tompkins, & Rau, 1988; Schulz, Tompkins, Wood, & Decker, 1987). On the whole, these models provide a convenient framework for organizing the large number of variables relevant to understanding the caregiving process.

The Basic Stress-Coping Model

Probably the most fundamental way to conceptualize the caregiving experience is in terms of a framework for interactions between the individual and the environment (Elliott & Eisdorfer, 1982). This model has three primary elements: a potential activator (x), an individual's reaction (y) to the activator, and the consequences (z) or sequelae to the reactions. Mediators are thought to be the filters and modifiers that act on each stage of the x-y-z sequence to produce individual variations. In the laboratory, specifying the x-y-z sequence can be relatively straightforward. For example, injecting an antigenic substance (x) under the skin of a healthy individual results in an immunologic response (y) that produces local swelling, redness, and tenderness (z) (Elliott & Eisdorfer, 1982). Nonlaboratory situations, however, are not as easy to characterize or understand. A given activator may elicit a strong reaction in one person and none at all in another, or it may result in a response at one point in time but not another. Moreover, distinctions between reactions and consequences are often difficult to make.

The basic x-y-z model, elaborated and applied to many caregiving situations, has been very useful in identifying and organizing variables thought to affect caregiving outcomes (see Biegel, Sales, & Schulz, 1991 for a review of stress models applied to caregiving). One of the most recent iterations of a stress process model may be particularly applicable to caregiving in linking environmental stressors to health outcomes (Cohen, Kessler, & Gordon, 1995). An adaptation of this model is presented in Fig. 8.1.

The sequential relations between components of this model can be described as follows. The primary stressors or environmental demands include the functional limitations and problem behaviors of the disabled individual along with related social and environmental stressors. When confronted with these stressors, people evaluate whether the demands pose a potential threat, and whether sufficient adaptive capacities are available to cope with them. If they perceive the environmental demands as threatening and at the same time view their coping resources as inadequate, they perceive themselves as under stress. The appraisal of stress is presumed to result in negative affect, which under extreme conditions may contribute directly to the onset of affective psychiatric disorders. Negative emotional responses also may trigger behavioral or physiological responses that place the individual at increased risk for psychiatric or physical illness.

It also is conceivable, although less likely when applied to challenges such as dementia caregiving, that stressors are appraised as benign and/or that individuals feel they have the capacity to deal with the stressors. This, in turn, leads to positive emo-

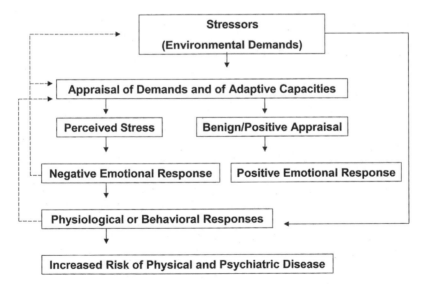

FIG. 8.1. The stress-health process applied to caregiving. (Adapted from "Measuring Stress" by S. Cohen, R. C. Kessler, and L. U. Gordon, 1995, Oxford University Press, New York)

tional responses that may lead to salutary physiologic and behavioral responses. Although this pathway is theoretically possible and has been demonstrated empirically in some instances (Beach, Schulz, Yee, & Jackson, 2000), it is important to note that this pathway is less common and, on the whole, has less empirical support.

Two other features of this model are important. One critical feature is that environmental stressors can place the individual at risk for negative health outcomes even when appraisal of the stressor does not result in perceptions of stress or negative emotional responses. This is illustrated in Fig. 8.1 by the arrow linking stressors directly to physiological or behavioral responses. For example, caregivers may take pride in doing an excellent job of caring for a demented relative without realizing that they are neglecting their own needs, such as eating regularly or seeing a doctor for their own health problems.

The second feature of this model concerns the existence of many possible feedback loops, a few of which are illustrated with dashed lines in Fig. 8.1. Although the model is primarily unidirectional, dealing with stressors is a complex, dynamic process in which responses at one stage of the model may subsequently feed back to earlier stages. One example of this process is represented by the dashed line linking emotional responses to the stressor and the appraisal process. A negative emotional response to a stressor might subsequently increase the stressor itself (e.g., cause the care recipient to increase problem behaviors) or impact negatively on appraisal of the stressor. This might happen, for example, when a caregiver becomes distressed in response to the disruptive behavior of a care recipient, who then becomes more disruptive because of the caregiver's response.

Enduring Outcomes—Caregiving End Points

Caregiving end points are the prolonged or cumulative consequences of being exposed to the demands of caregiving. The stress–health model presented here focuses on health as the primary outcome of the stress–health process because there is strong consensus that health is an important outcome at both the individual and societal level, and because stress and health have been linked consistently in the empirical literature. In addition, goal of most interventions for caregivers is to improve or maintain the psychological and physical well-being of the caregiver. Nevertheless, this focus on health should be viewed as illustrative rather than definitive. A wide range of caregiving effects have been described in the literature including disruption of family routines, financial hardship, work-related problems, psychological distress, and psychological and physical morbidity, including mortality. Feeling burdened or distressed by the demands of caregiving is the most frequently reported outcome associated with caregiving, although this is not a universal phenomenon, particularly among spousal caregivers (Schulz et al., 1995). Psychiatric morbidity, such as depression and anxiety, also is common. Physical health effects, such as increased susceptibility to illness, have been more difficult to demonstrate, although they are likely to occur in high-demand situations among vulnerable (e.g., frail) caregivers. A recent study by Schulz and Beach (1999) showed that spousal caregivers who report caregiving strain are at increased risk of mortality. Possible mediators of the mortality effect include depression associated with caregiving and changes in health-related behaviors (e.g., sleeping and eating patterns, medical compliance).

Caregiving is not always stressful, particularly among spousal caregivers in the early stages of a caregiving career. Although the available epidemiologic data are limited, the work of Schulz and Beach (1999) suggests that in older married dyads wherein at least one spouse has a functional disability, approximately 80% provide care. Furthermore, among those with a disabled spouse, approximately 45% provide care and report strain associated with caregiving, whereas 35% provide care and report no strain. Positive effects of caregiving such as increased self-esteem, the satisfaction of knowing that one's relative is receiving proper care, and improved mental health, also have been reported in the literature (Beach et al., 2000). In general, positive effects are more likely to occur in the early stages of caregiving when the stressors are mild to moderate. Anecdotal evidence also suggests that some caregivers benefit from the increased control and responsibility afforded by caregiving. Finally, given the complexity of caregiving roles, it is not uncommon for caregivers to experience positive and negative outcomes simultaneously. Thus, a caregiver may feel stressed and burdened on the one hand and experience enhanced self-esteem on the other.

Contextual Variables/Mediators and Moderators

The search for mediators and moderators of caregiving outcomes is motivated in part by the lack of a strong relation between objective characteristics of the stressor and caregiver outcomes. For example, functional disability of the care recipient is only moderately related to caregiver depression. Therefore, other factors need to be considered in linking independent with dependent variables.

Although not illustrated in Fig. 8.1, all models of caregiving recognize that contextual or situational variables and individual differences contribute to caregiving outcomes. This category of variables is broadly defined to include the social networks and support systems of caregivers; the characteristics of caregivers including socioeconomic status, health, gender, and relationship to patients; the quality of the relationship between caregiver and care recipient; the number of competing roles such as mother, wife, and worker; and personality attributes such as orientation toward control and neuroticism. It also includes factors characterizing the environment, such as the availability and use of professional services.

Studies focused on these variables have yielded a significant body of reliable findings (Horowitz, 1985; Schulz et al., 1995). At the descriptive level, investigators have characterized the caregiving population in terms of gender, age, race, ethnicity, marital status, employment, economic status, health status, and living arrangements. For example, it is known that (a) most caregivers are women; (b) their average age is approximately 57 years; (c) about 70% of all caregivers are married; (d) one third of informal caregivers are employed, although as a group, both male and female caregivers are less likely to be employed than similarly aged counterparts; (e) compared with their age peers in the general population, male and female caregivers are more likely to report adjusted family incomes below the poverty level; (f) the self-assessed health of caregivers is lower than that of their age peers; and (g) approximately three fourths of caregivers live with the disabled family member or friend (Ory, Hoffman, Yee, Tennstedt, & Schulz, 1999).

A second body of research examines these variables in terms of their direct relation to caregiving impact. As would be expected, living arrangements between caregivers and care recipients are major predictors of caregiver involvement, behavior, and burden. Caregivers who live with the impaired elderly are more involved with the daily care of patients and experience greater limitations on their personal lives. Employed caregivers frequently experience conflict between the demands of work and the needs of their care recipient. Caregivers with a great deal of social support cope better with the demands of caregiving than those with little support.

A third body of research treats these variables as interactive conditioning factors that moderate relations between stressors and their impact on caregivers. One example of this approach is the stress-buffering hypothesis applied to social support. According to this view, individuals exposed to high levels of caregiving stress benefit from support given by others, but individuals who are not stressed or who experience low levels of stress as a result of caregiving exhibit no beneficial effects attributable to social support. One of the important contributions of research on social support is its emphasis on identifying mechanisms through which caregiving stressors exert their impact on caregiver outcomes. For example, support may play a role at two different points in the causal chain linking stress to illness. It may intervene between the stressor and a stress reaction by attenuating or preventing a stress appraisal response, or it may intervene after the stress is experienced to prevent the onset of pathologic outcomes by reducing emotional reactions, dampening physiologic processes, or altering maladaptive behavior responses. Just knowing that others are available to help care for patients when necessary may prevent caregivers from feeling burdened or stressed. Receiving support may dampen the impact of a perceived stressor by providing helpful information or assistance or by facilitating healthful behaviors.

Finally, it is worth noting that well-known vulnerabilities for negative mental health outcomes also apply to caregivers. Individuals low in socioeconomic status, compromised in health status, or with a history of mental health problems are at higher risk for negative outcomes than caregivers who do not fall into these categories (Schulz et al., 1995).

Clearly, a comprehensive model of the caregiving experience would be much more complex than the stress-health process illustrated in Fig. 8.1. This complexity represents both a challenge and an opportunity. On the one hand, researchers are never likely to understand fully the caregiving experience and its many individual variations. On the other hand, the rich and interactive nature of the caregiving experience presents opportunities for the creative interventionist interested in enhancing the caregiver's quality of life.

CAREGIVING INTERVENTION RESEARCH

Although anecdotal reports of early intervention efforts were generally positive, the first critical reviews of the literature were considerably more sobering. Toseland and Rossiter's (1989) review of 29 studies concluded that, although caregivers evaluated interventions positively, there was "no clear link ... between participants' satisfaction (with group interventions) and other important outcomes for caregivers, such as improving coping skills, preventing psychological disturbances, increasing caregiver support systems, or improving caregivers' ability to care for themselves" (p. 438). Similarly, "time-limited psychoeducational interventions have modest therapeutic benefits as measured by global ratings of well-being, mood, stress, psychological status, and caregiving burden" (p. 481). Focusing exclusively on interventions aimed at alleviating caregiver distress, Knight, Lutzky, and Macofsky-Urban (1993) concluded that individual psychosocial interventions and respite programs are moderately effective, though psychosocial interventions with groups are less effective. Zarit and Teri (1992), in describing the available intervention literature as the "first generation" of studies, pointed out that interpretation of this preliminary work should be tempered by the fact that expectations for particular intervention outcomes and the malleability of caregivers have been overly optimistic, and that some intervention effects may be underestimated because of methodologic limitations of the studies.

Bourgeois, Schulz, and Burgio (1996) organized their review of the literature around six broad categories: support groups, individual and/or family counseling, case management, respite and day-care services, skills training, and various combinations of these strategies. Several general conclusions can be derived from their analysis of the literature. First, the complexity and rigor of intervention studies continue to improve with an increasing emphasis on randomized designs. Second, the literature on the whole supports the conclusion that more is better. Multicomponent interventions that blanket caregivers with a diversity of services and supports, hoping that a combination of components will have an impact on a caregiver's unique needs, tend to generate larger effects than narrowly focused interventions. Similarly, single-component interventions with higher intensity (frequency and duration) also have a greater positive impact on the caregiver than similar interventions with lower intensity. Third, achieving generalization of effects beyond the specific target of an intervention has been difficult. For example, a skills training intervention may effectively enhance caregivers' ability to manage care recipients, but may not necessarily reduce their sense of subjective burden.

Given the complexity of the caregiving experience, the variability in caregiver resources, and the variety of outcomes examined, it should not be surprising to find that more recent reviews of the caregiving intervention literature also have been unable to identify a silver bullet solution that will alleviate caregiver distress (Kennet, Burgio, & Schulz, 2000; Schulz, 2000). There is no single, easily implemented, and consistently effective method for eliminating the stresses of caregiving.

Virtually all interventions studies examined in the recent review by Kennet et al. (2000) reported some level of success and, as a group, they provide valuable insights into different methods for achieving caregiver impact as well as the pitfalls of conducting intervention research in this complex area. There exists a strong consensus that all caregivers are likely to benefit from enhanced knowledge about the disease, the caregiving role, and resources available to caregivers. Once the informational needs have been met, caregivers may benefit additionally from training in general problem-solving skills as well as in more specific skills for managing patient behaviors or their own emotional response to caregiving. Some studies currently underway teach the caregiver rudimentary behavior modification skills, including behavior assessment techniques and methods for changing undesirable behaviors, such as repetitive questions and wandering. Caregivers are taught to identify possible eliciting conditions for undesirable behaviors and then to change those conditions systematically to eliminate or reduce the unwanted behavior. Recent intervention studies also suggest that there may be important synergies achieved by simultaneously treating the care recipient (e.g., giving medications or memory retraining) and the caregiver. Suggestive evidence for this hypothesis is reported by Shikiar et al. (2000), who showed that an experimental cognitive-enhancing drug for the patient combined with limited education and support for the caregiver resulted in a significant reduction in caregiver distress as compared with the group receiving a placebo.

The existing literature also points to a rich array of methods for delivering interventions to caregivers. Among these are traditional methods, such as individual and group sessions, but also newer technologies involving enhanced telephone systems, microcomputers, and Web sites. As sophisticated communication technologies become more available to individuals, treatment delivery options will increase. Although new technology has the potential to overwhelm already stressed caregivers, it can be effective if introduced in a graduated stepwise fashion (see Czaja, chap. 9, this volume).

Many challenges remain for caregiving intervention researchers, but two deserve special emphasis. One concerns the choice of outcomes the other has to do with methods for identifying the optimal mix of intervention components for a particular caregiver. Existing studies focus on a wide range of outcomes, including caregiver distress, physical and mental health, care recipient behavior, and care recipient institutionalization. In selecting an appropriate outcome for an intervention, the authors think it useful to identify clearly the proximal and distal goals of the intervention. Proximal outcomes are those on which the intervention is directly intended to have an effect, whereas distal outcomes are typically contingent on first achieving the proximal outcomes. As an example, the proximal outcome for an intervention aimed at changing patient problem behaviors would be a reduction in unwanted behaviors. A distal outcome might include a reduction in caregiver distress. Too often, interventions fail to assess relevant proximal outcomes, making it difficult to understand how or why an intervention did or did not work. In addition, interventionists frequently focus on distal outcomes (e.g., time to institutionalization) without considering the relation between proximal goals of the intervention (e.g., enhancement of knowledge about the disease, professional support options) and the distal outcomes. One useful approach to achieving consistency between goals and intervention methods is to carry out a detailed task analysis of the intervention. This exercise forces the researcher to articulate specific goals and subgoals as well as their relation to specific treatment components of the intervention.

A recurrent theme in this literature is that caregivers have multiple needs that interventionists must address to maximize impact. Finding the optimal mix of program elements for a given caregiver–care recipient dyad at a particular point in the disease trajectory should be a major goal of intervention researchers. However, achieving this goal is virtually impossible with studies of limited sample size and limited intervention approaches. What is needed are large-scale studies with diverse populations that enable the full exploration of complex interactions among caregiver and care recipient characteristics, treatment components, and methods of delivery. An important first step in this process is the development of a conceptual model that adequately captures the full range of possible intervention approaches.

DEVELOPING A CONCEPTUAL MODEL FOR CAREGIVER INTERVENTIONS

Attempts to characterize interventions used in caregiving studies have typically focused on the broad goals of the intervention (Bourgeois et al., 1996; Dunkin & Anderson-Hanley, 1998; Knight et al., 1993). For example, interventions have been classified in terms of the general functions they serve (e.g., education, support, or respite), but no attempt has been made to develop a detailed taxonomy of caregiver interventions. This section describes a comprehensive classification system for interventions that captures content, process, and goals of an intervention in a theoretically relevant manner.

The authors propose that interventions be characterized along three relatively orthogonal dimensions: (a) the targeted primary entity (i.e., the caregiver, care recipient, social and physical environments of the caregiver and care recipient), (b) the targeted primary functional domain (i.e., knowledge, cognitive skills, behavior, affect), and (c) the delivery characteristics of an intervention (i.e., method of delivery, intensity, adaptability and controllability of the intervention to individual needs). The first two dimensions are derived from the general stress–health model, whereas the third dimension captures pragmatics of the delivery system that may attenuate the impact of an intervention. Each of these dimensions is further defined in the following discussion and in the accompanying tables.

Targeted Primary Entity

The caregiver typically serves as the vehicle through which all interventions are delivered, but the interventions vary with respect to the locus of their primary intended effect. For some interventions, the primary goal is to change the physical or social environment, whereas for others it is to induce change (e.g., in behavior, affect) within the caregiver or the care recipient (Table 8.1)

Targeted Primary Functional Domain

Interventions also vary with respect to the primary targeted functional domain. For example, some interventions are designed primarily to increase the knowledge of the caregivers about themselves, the care recipient, or the social environment (Table 8.1). Other interventions emphasize more generalizable problem-solving skills that could be applied to many different situations. The goal of still other interventions might be to enhance the behavior skills of the caregiver. Finally, some interventions are intended to alter specific emotional responses to caregiving directly, such as anger or sadness.

TABLE 8.1

Target Entity/Functional Domain Matrix Applied to Caregiving

Functional Domain Definition	Target Entity		
	Caregiver	Care Recipient	Social/Physical Environment
Cognition Knowledge: An intervention that involves provision of knowledge (written or video).	Information and knowledge related to caregiving that concerns primarily the caregiver.	Information and knowledge about the care recipient that might be useful to the caregiver.	Information on how family members other than the caregiver respond to caregiving; information about characteristics of the physical environment that might impede or facilitate caregiving.
Cognition Skills: An intervention that involves teaching an approach or technique that enhances caregiving.	Cognitive behavior therapy (CBT) and problem solving skills training applied to the caregiver typify this cell.	Caregiver learns antecedents of care recipient's problem behaviors and methods for dealing with them, and may acquire skills for enhancing cognitive functiooning of the care recipient.	Cognitive coping skills aimed at helping the caregiver deal with negative attitudes and interpersonal/communication skills aimed at generating positive attitudes of social environment.
Behavior: An intervention that actively engages the caregiver in a behavioral strategy (e.g., role playing).	Skills to better monitor and regulate the caregiver's own behavior in relation to the care recipient and other persons are demonstrated and practiced.	Skills to better monitor and regulate the behavior of the care recipient are demonstrated and practiced.	Practice of behavioral skills aimed at changing behavior of persons in the social environment such as other family members; demonstration/ practice on how to assess the physical environment .
Affect: An intervention designed to directly impact on the affect/emotion of the target entity.	Caregiver learns to identify and engage in pleasant activities.	Caregiver learns to provide pleasant diversion when care recipient becomes distressed.	(Limited to social environment). Caregiver learns strategies that maximize support from the social environment.

Delivery Characteristics

In addition to the target and functional domain of an intervention, it is important to describe its delivery characteristics. Three types of delivery characteristics can be identified (Table 8.2). The first and most common delivery attribute is intensity or dose. This dimension is analogous to the dosage of medication that might be administered to individuals in an experimental drug trial. When applied to psychosocial interventions, it includes three factors: the frequency of contact between intervention agent and caregiver, the duration of the contacts, and the overall duration of the intervention. In general, it is believed that higher doses lead to better outcomes, although this need not always be the case.

The second general delivery characteristic is the method of delivery. This category includes a broad array of attributes that describe the way an intervention is delivered to the participant. Included are factors such as whether the intervention is delivered at the individual or group level, and whether it occurs face-to-face or via technological devices such as a telephones or microcomputers. Relatively little is known about how these delivery characteristics affect treatment outcomes, although it seems reasonable to assume that delivery mode would interact with the intended functional domain of an intervention. Therefore, if the primary goal of an intervention is to enhance caregiver knowledge, group sessions or microcomputers could be an effective means for achieving this. On the other hand, if the goal of an intervention is to enhance the behavior management skills of the caregiver, a one-on-one interaction in which the caregiver has the opportunity to try out and demonstrate acquired skills is likely to be more effective.

As the technology for delivering interventions of all type expands, it will become increasingly important to assess the impact of delivery methods on treatment outcomes. The availability of new communication technologies has the potential to increase the cost effectiveness of intervention efforts.

The final delivery attribute concerns the adaptability or controllability of the intervention by either the interventionist or the study participant. This important dimension has received relatively little attention in the literature. For example, many psychosocial interventions allow the study participant to identify which problems will be addressed and what method will be used to address them, as well as when and how often the intervention will be delivered. The authors would characterize such an intervention as highly individualized to the needs of the participant, predicting that it is more likely to achieve desired outcomes than an intervention fixed in terms of what, how, and when it is delivered to the participant. Interventions also may vary in the extent to which they systematically solicit participant input on their structure and content. Finally, at a more global level, interventions may vary in the extent to which they permit changes in the protocol once a study is underway. Medical intervention trials often are constrained in this respect once the protocol has been implemented, but psychosocial interventions often permit significant adjustments in the protocol even after a study has begun. Although it is beyond the scope of this chapter to argue the advantages and disadvantages of these varying approaches to the delivery of an intervention, no one would argue with the suggestion that interventions should be clearly described on each of these characteristics.

Theoretically, all three types of intervention characteristics (target, functional domain, and delivery characteristics) can be orthogonal to each other. Thus, one could create a three-dimensional space and locate an intervention or components of an intervention in that space. Given the complexity of thinking about different interventions in terms of all three dimensions simultaneously with multiple factors in each dimension, it is convenient to characterize interventions first in terms of two dimensions, the entity and targeted domain, and then separately consider the delivery system used to implement the intervention. In the caregiving context, combining the entity and domain dimensions yields a 3 (entities) by 4 (domains) matrix. Definitions and examples for each cell of this matrix are provided in Table 8.1. Delivery system characteristics that define dose, mode of delivery, and adaptability/controllability are listed in Table 8.2.

Elements of this classification system can be linked to the stress–process model in the following way. The authors would argue that interventions targeting the care recipient or the social and physical environment represent attempts to alter environmental stressors. For example, a behavior skills training intervention that trains the caregiver how to identify and change antecedent conditions that lead to disruptive behaviors in the care recipient would be classified as a skills training intervention targeting the care recipient. As shown in Fig. 8.2, such interventions target a primary source of stress in the sequential model. Similarly, interventions targeting caregiver cognitions about their abilities as caregivers should have their primary impact on the appraisal of demands and adaptive capacities. Interventions targeting caregiver affect, such as feelings of depression and anxiety, are aimed at altering emotional responses of the caregiver. Thus, interventions can be conceptualized as primarily targeting a specific component of the stress–health sequential model. Because of the multiple feedback loops in the sequential model, an intervention targeting one component of the model should, over time, affect other components as well.

This approach to conceptualizing interventions has several advantages. First, it provides a common framework for characterizing many different approaches to interven-

TABLE 8.2

Delivery Charateristics of Psychosocial

Intensity	Method of Delivery	Adaptabilty/Controllability
Frequency of contact	Type of contact (one on one, group, family)	Protocol modification allowed
Duration of Contact	Mode of contact (in person, telephone, computer)	Interventionist modifications as specified by protocol
Duration of intervention period	Location (home, classroom, lab)	Study participant modifications allowed
	Mechanism of delivery (written, verbal, role play, video, computer)	

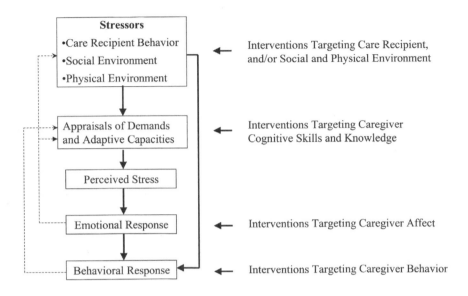

FIG. 8.2. Intervention strategies and their role in a stress model.

tions for caregivers. This should facilitate comparisons across studies and enable meta-analyses of caregiver intervention research. Second, identifying discrete components of multifaceted interventions may help researchers to identify key features of an intervention that contribute to positive outcomes. Because many interventions target multiple entities and domains simultaneously, the ability to partition them across multiple cells of the matrix should enhance the ability to ascribe causality. Finally, the inclusion of delivery system characteristics, which are often overlooked and rarely measured systematically in caregiver intervention research, helps researchers to explore how functional domains and delivery methods interact to produce desired outcomes. In general, it would be expected that interventions high in multiple dimensions (i.e., multiple domains and targets, at high doses, with high degrees of adaptability/controllability) would be more effective than interventions low in these dimensions.

SUMMARY AND CONCLUSION

The goal in this chapter is to provide a framework for thinking about the caregiving experience and for characterizing intervention approaches used with caregivers. Although the discussion is not exhaustive, it does describe the dominant themes and issues of concern to researchers and practitioners. No doubt new perspectives and even more challenging issues will emerge as research in this area progresses.

Because emerging computer and communications technology will further increase the range of options in delivering caregiver interventions, it is all the more important that a taxonomy is in place for characterizing psychosocial interventions. An important secondary goal of this chapter is to describe such a taxonomy by focusing on three elements characteristic of all interventions: target, functional domain, and delivery characteristics. Together, these characteristics define a wide range of attributes that describe all psychosocial interventions. Their systematic application in future studies should help us in better assessing and understanding the impact of complex psychosocial interventions.

ACKNOWLEDGMENTS

This chapter was prepared for the conference on Human Factors Interventions for the Health Care of Older Adults sponsored by the Center for Applied Cognitive Research on Aging, February 23–27, 2000, Destin, Florida. Preparation of this chapter was supported in part by grants from the National Institute of Mental Health (R01 MH 46015, R01 MH52247, T32 MH19986), the National Institute on Aging (AG13305, AG01532), and the National Heart, Lung, and Blood Institute (P50 HL65112).

REFERENCES

Beach, S. R., Schulz, R., Yee, J. L., & Jackson, S. (2000). Negative and positive health effects of caring for a disabled spouse: Longitudinal findings from the Caregiver Health Effects Study. *Psychology and Aging, 15,* 259–271.

Biegel, D., Sales, E., & Schulz, R. (1991). *Family caregiving in chronic illness: Heart disease, cancer, stroke, Alzheimer's disease, and chronic mental illness.* Newbury Park, CA: Sage.

Biegel, D. E., & Schulz, R., (1999). Caregiving and caregiver interventions in aging and mental illness. *Family Relations, 48,* 345–354.

Bourgeois, M. S., Schulz, R., & Burgio, L. (1996). Interventions for caregivers of patients with Alzheimer's Disease: A review and analysis of content, process, and outcomes. *International Journal of Aging and Human Development, 43,* 35–92.

Cohen, S., Kessler, R. C., & Gordon, L. U. (1995). *Measuring stress.* New York: Oxford University Press.

Cohler, B., Groves, L., Borden, W., & Lazarus, L. (1989). Caring for family members with Alzheimer's disease. In E. Light & B. Lebowitz (Eds.), *Alzheimer's disease treatment and family stress: Directions for research* (pp. 50–105). Washington, DC: National Institute of Mental Health.

Dunkin, J. J., & Anderson-Hanley, C. (1998). Dementia caregiver burden: A review of the literature and guidelines for assessment and intervention. *Neurology, 51(Suppl 1),* S53–S60.

Elliot, G. R., & Eisdorfer, C. (1982). *Stress and human health.* New York: Springer.

Haley, W. E., Levine, E. G., Brown, S. L., & Bartolucci, A. A. (1987). Stress, appraisal, coping, and social support as predictors of adaptational outcome among dementia caregivers. *Psychology and Aging, 2,* 323–330.

Horowitz, A. (1985). Family caregiving to the frail elderly. In M. P. Lawton & G. Maddox (Eds.). *Annual review of gerontology and geriatric.* (Vol. 5, pp. 194–246). New York: Springer.

Kennet, J., Burgio, L., & Schulz, R. (2000). Interventions for in-home caregivers: A review of research 1990 to present. In R. Schulz (Ed.), *Handbook on dementia caregiving: Evidence-based interventions for family caregivers* (pp. 61–126). New York: Springer.

Knight, B. G., Lutzky, S. M., & Macofsky-Urban, F. (1993). A meta-analytic review of interventions for caregiver distress. *Gerontologist, 33,* 240–248.

Kramarow, E., Lentzner, H., Rooks, R. Weeks, J., & Saydah, S. (1999). *Health and aging chartbook. Health, United States, 1999.* Hyattsville, MD: National Center for Health Statistics.

Montgomery, R. J. V., Stull, D. E., & Borgatta, E. F. (1985). Measurement and the analysis of burden. *Research on Aging, 7,* 137–152.

National Alliance for Caregiving and the American Association of Retired Persons. (1997). *Family caregiving in the U.S.: Findings from a national survey. Final Report.* Bethesda, MD: National Alliance for Caregiving.

Ory, M. G., Hoffman III, R. R., Yee, J. L., Tennstedt, S., & Schulz, R. (1999). Prevalence and impact of caregiving: A detailed comparison between dementia and nondementia caregivers: Dementia and nondementia caregiving. *The Gerontologist, 39,* 177–185.

Schulz, R. (Ed.). (2000). *Handbook on dementia caregiving: Evidence-based interventions for family caregivers.* New York: Springer.

Schulz, R., & Beach, S. (1999). Caregiving as a risk factor for mortality: The caregiver health effects study. *Journal of the American Medical Association, 282,* 2215–2219.

Schulz, R., O'Brien, A, T., Bookwala, J., & Fleissner, K. (1995). Psychiatric and physical morbidity effects of Alzheimer's Disease caregiving: Prevalence, correlates, and causes. *The Gerontologist, 35,* 771–791.

Schulz, R., Tompkins, C. A., & Rau, M. T. (1988). A longitudinal study of the psychosocial impact of stroke on primary support persons. *Psychology and Aging, 3,* 131–141.

Schulz, R., Tompkins, C. A., Wood, D., & Decker, S. (1987). The social psychology of caregiving: Physical and psychological costs to providing support to the disabled. *Journal of Applied Social Psychology, 17,* 401–428.

Shikiar, R., Shakespeare, A., Sagnier, P-P. Wilkinson, D., McKeith, I., Dartigues, J-F. & Bubois, B. (2000). The impact of Metrifonate therapy on caregivers of patients with Alzheimer's disease: Results from the MALT clinical trial. *Journal of the American Geriatrics Society, 48,* 268–274.

Toseland, R. W., & Rossiter, C. M. (1989). Group interventions to support family caregivers: A review and analysis. *The Gerontologist, 29,* 438–448.

Zarit, S. H., & Teri, L. (1992). Interventions and services for family caregivers. In K. W. Schaie & M. Powell Lawton (Eds.), *Annual review of gerontology and geriatrics.* (Vol. 11, pp. 287–310). New York: Springer.

9

Telecommunication Technology as an Aid to Family Caregivers

Sara J. Czaja
University of Miami School of Medicine

Over the past several decades a number of demographic trends, such as the increased number of older people in the population, have increased the need for informal caregiving. Generally, the prevalence of chronic conditions or illnesses such as dementia, diabetes, heart disease, or stroke increase with age. Consequently, older adults (especially the "oldest old") are more likely to need some form of care or assistance. Approximately 7 million people older than 65 years have mobility or self-care limitations (Fig. 9.1) and about 4 million Americans suffer from Alzheimer's disease (AD). The prevalence of AD is expected to increase with the aging of the population. In fact, the prevalence rates for dementia and AD double approximately every 5 years beyond the age of 65 years (Fig. 9.2). Most people afflicted with AD live at home, where they are cared for by family members.

Advances in medical technology also have made it possible for people with special needs, such as the developmentally disabled, to live longer. These individuals also require care and assistance with routine tasks. Typically, this assistance is provided by family members. Currently, about 15% of U.S. adults are providing care for a seriously ill or disabled relative (Otten, 1991). (See chap. 8, this volume, for a more complete discussion of caregiving.)

Caregiving poses a range of physical, emotional, and financial demands and often causes burden and stress for family caregivers. The negative impact of caregiving on family members is well documented and includes anxiety, depression, financial strain, adverse health outcomes and disruption in family dynamics (Schulz & Beach, 1999; Schulz, O'Brien, Bookwala & Flessner, 1995). Given the large number of people who will need some form so they can live in the community, there

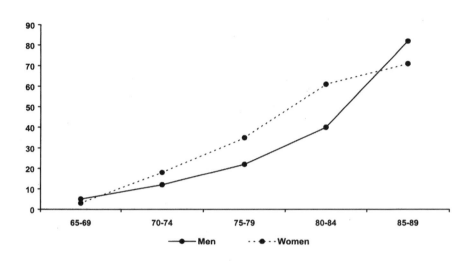

FIG. 9.1. Age- and gender-specific incidence rates of Alzheimer's disease. (From "Incidence of Dementia and Probably Alzheimer's Disease in a General Population," by D. L. Bachman et al., 1993, *Neurology, 43*, 515–519.) Copyright © 1993 Lippincott, Williams and Wilkens. Reprinted with permission.

is a tremendous need to develop interventions that will improve the well-being and quality of life of both caregivers and patients.

Caregivers represent an extremely important resource both to their families and the community at large. It would be extremely costly to replace informal caregivers with paid care. Recent estimates indicate that in 1997, the economic cost of informal caregiving was equivalent to approximately 18% of the total national spending for health care (Paveza et al., 1998). The need to develop interventions for caregivers is underscored by current demographic trends such as greater numbers of women in the labor force, more older adults living alone, and fewer children available to provide care. By the year 2030, the average number of children per family will be approximately 2 as compared with 3 in 1990, and the number of married couples with no children is expected to increase to 7 million by the year 2010 (U.S. Bureau of the Census, 1997). These trends suggest that there will be fewer family caregivers available to provide care to disabled family members. In fact, it is estimated that there will be approximately a threefold increase in the parent support ratio (i.e., the number of persons older than 85 years per 100 persons ages 50 to 64 years) over the next several years.

During the 1990s, a number of studies addressed interventions for family caregivers. These interventions include community and family support groups, respite care programs, and psychoeducation programs such as skills training (Kennet,

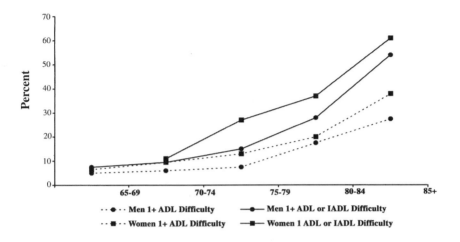

FIG. 9.2. Prevalence of difficulties with activities of daily living (ADL) and instrument activity of daily living (IADL) among community elderly, 1985. From *Essentials of Clinical Geriatrics* (p. 30), by R. L. Kane, J. G. Ouslander, and I. B. Abrass (Eds.), 1994, New York: McGraw Hill. Copyright © by McGraw Hill. Reprinted with permission.

Burgio, & Schulz, 2000). However, despite the proliferation of these intervention programs, they have met with only limited success for a variety of reasons. Services are not always available to caregivers, and many caregivers are unwilling to use available community services because of issues such as cost, logistic problems, or feelings of guilt about receiving help from outside the family structure. For example, problems such as difficulty arranging alternative help, inaccessible meeting places, or scheduling conflicts often prevent family caregivers from attending support group meetings (Wright, Lund, Pett, & Caserta, 1987). These data suggest that current intervention strategies may need to be modified or augmented if they are to respond more effectively to the needs of caregivers.

Current information technologies offer the potential of providing support and delivering services to caregivers and other family members. Computer networks can link caregivers to each other, health care professionals, community service, and education programs. Information technology also can enhance a caregiver's ability to access health-related information or information regarding community resources. Technology also can be used to facilitate the ability of health care providers to monitor the status of caregivers or care recipients or actually to deliver needed care.

The use of information technology as an intervention for caregivers is quite plausible given the increased use of technology within most domains and across most segments of the population. Currently, the number of Internet users in the United States is approximately 79 million, which represents an increase of 36% since 1997 (McPhee, 1998). Furthermore, the number of publicly available databases grew from 400 in 1980 to 8,400 in 1993, and the number of online services rose from 59 to

825 during this same period (Marchioni, 1995). Most people, even older people, are willing to use technology for tasks such as information seeking, communication, education, or shopping. For example, (Czaja, Guerrier, Nair, and Landauer (1993) found that a sample of community-dwelling older women, ranging in age from 55 to 95 years, were able to use an electronic message system to perform communication tasks. Moreover, the participants reported that they enjoyed using the system and found it valuable to have a computer in their home. Specifically, the participants reported that they liked using the system because it allowed them to communicate with others short of leaving their homes and gave them a chance to meet new people and learn something new.

Although the number of people older than 55 years who use the Internet is low compared with other age groups, usage among this age group is increasing (Fig. 9.3). Currently, about 30% of people ages 55 to 75 years own a personal computer, an increase of 21% since 1994 (Teel, 1999). Of people older than 50 years, 17% are Internet users, and the number of users in this age group is increasing at the same rate as the overall Internet population. Of course, as discussed by Stronge, Walker, and Rogers (chap. 14, this volume), designers must address a number of interface issues to help ensure that older people can successfully adapt to technological developments such as the Internet. To demonstrate the feasibility of using technology as an intervention strategy, examples of research projects in this area are provided. The final section of this chapter identifies areas in which more research is needed.

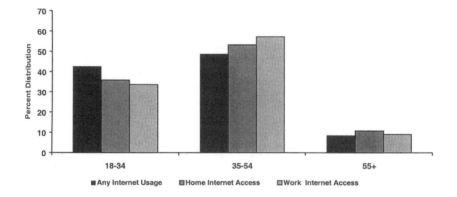

FIG. 9.3.   Adults older than 55 years using Internet. (From Mediamark Research Inc., New York, CyberStats, Spring 1998. http://www.mediamart.com/pages/freedata.htm.)

THE POTENTIAL ROLE OF INFORMATION TECHNOLOGY AS AN AID TO FAMILY CAREGIVERS

Although the negative consequences associated with caregiving are well documented, it is difficult to develop effective intervention strategies because the contextual demands of caregiving are so complex. For example, the caregiver's response to caregiving demands varies as a function of age, gender, ethnicity, and the relationship of the caregiver to the care recipient (e.g., daughter vs. spouse) (Ory, Yee, Tennstedt, & Schulz, 2000). Caregiving demands also vary with the nature and time course of the patient's illness (Czaja, Eisdorfer, & Schulz, 2000). For example, in the presymptomatic stages of the disease, caregivers may need education about the disease and advice regarding life planning. However, in the middle to late stages, they may need in-home services and counseling in addition to education.

Many caregivers have limited knowledge about the patient's illness and symptoms, and thus do not have realistic expectations regarding the patient's functional status or behavior responses as the disease progresses. Furthermore, many lack the skills needed to meet caregiving demands. Also, many caregivers are unaware of available formal support services, and thus make limited use of these services. Finally, many caregivers are isolated and lose contact with family and friends. The literature suggests generally that caregivers with social support from family and friends cope better with the demands of caregiving than those with little support (Schulz, Gallagher-Thompson, Haley, & Czaja, 2000).

Therefore, a number of challenges face intervention researchers including the need to develop strategies to increase caregivers' knowledge about the care recipient's disability/disease and associated behavior symptoms; to increase caregivers' knowledge and use of available resources and services; to enhance communication between caregivers and health care professionals, family members, friends, and other caregivers; and to enhance the skill level of caregivers so they are better able to cope with the physical and emotional demands associated with caregiving tasks.

Information technology may be used to help meet the challenges associated with caregiving and offers the potential to improve the quality of life for both caregivers and care recipients. As shown in Table 9.1, a variety of technologies and technological applications have proved to be effective for this population. Computer-based communication can be used to provide support and reduce the isolation of caregivers by facilitating links to friends, family members, other caregivers, and local support organizations. In contrast to face-to-face support groups, electronically-based support groups are more convenient because they do not require caregivers to leave home or disrupt caregiving routines. Computer communication also can occur asynchronously, which is beneficial for family members who live in different time zones or who have alternative work schedules such as shift work. Computer networks can be accessed at any time of the day or night. This is especially beneficial in times of crisis.

Gallienne, Moore, and Brennan (1993) found that access to a computer network, ComputerLink, increased the amount of psychological support provided by nurses to homebound caregivers of Alzheimer's patients. ComputerLink also enabled caregivers to access a caregiver support network that enabled them to share

TABLE 9.1

Example of Technology as an Intervention
for Caregivers and Patients

Investigators	Technology/Application	Target Population	Findings
Czaja, Guerrier, Nair, & Landauer, 1993	Simplified personal computer with e-mail, information services	Older women ranging in age from 50 to 95 years	System widely accepted and widely used
Gallienne, Moore, & Brennan, 1993	Computer network: Computer Link	Family caregivers of Alzheimers patients	Enhanced psychological support to caregivers; enhanced communication with other caregivers
Leirer, Morrow, Tanke, & Parienta, 1991	Voice mail	Elderly people with chronic conditions	Increased medication compliance and appointment attendance
Lund, Hill, Caserta, & Wright, 1995	Videotape respite program	Elderly patients with Alzheimer's disease	Provided relief for caregivers; reduced behavior agitation in patients
Wright, Bennet, & Gramling, 1998	Telecommunication system/ psychotherapy	Family caregivers of Alzheimer patients	Preliminary results: system well received by caregivers
Mahoney, Tennstedt, Friedman, & Heeren, 1999	Telecommunication system: TLC-Elder Care	Community-dwelling older adults with chronic conditions	System useful for monitoring health status; cannot completely substitute for face-to-face
Czaja, Rubert, & Eisdorfer (ongoing)	Computer integrated telephone system	Family caregivers of Alzheimer patients	System well received; online support groups popular
Mahoney et al., (ongoing)	Automated telecommunications system: TLC-AD	Family caregivers of Alzheimer patients	

experiences, foster new friendships, and gather information on the symptoms of the disease. Smyth and Harris (1993) are evaluating the use of a telecomputing information and support system for caregivers. The system allows caregivers to access information about dementing disorders and to communicate with one another via a caregiver forum.

Computer links also may facilitate communication between caregivers and long-distant relatives. It is quite common in the United States for family members to be dispersed among different geographic regions. In fact, nearly 7 million Americans are long-distance caregivers for older relatives (Family Caregiver Alliance, 1997). Clearly, network linkages can make it easier for family members to communicate, especially those who live in different time zones.

Technology also can aid caregivers' ability to manage their own health care needs as well as those of the patient by giving them access to information about medical problems, treatments, and prevention strategies. Software is available on several health-related topics such as stress management, caregiving strategies, and nutrition. A number of Web sites that might be useful to caregivers, including: The Mayo Health Oasis (www.mayohealth.org), Dr.Koop.com (www.drkoop.com), Americas Doctor (www.americasdoctor.com), and InteliHealth (www.intelihealth.com) from Johns Hopkins University. The Alzheimer's Association also has a home page on the World Wide Web (www.alzheimers.com). Electronic linkages also can enhance interactions between caregivers and healthcare providers. For example, caregivers may use applications such as e-mail to query health care professionals or communicate concerns about their patient's symptoms or specific problems that they are experiencing.

Conversely, information technology also may benefit health care providers by providing them easier access to clients and facilitating tasks such as daily health checks and reminders of home health care and medication regimens. Voice mail has been found to enhance compliance with medication and medical appointments among the elderly and persons with chronic conditions (Leirer, Morrow, Tanke, & Pariente, 1991; Morrow & Leirer, chap. 10, this volume). For example, TLC-Elder Care is a telecommunications system designed to automate monitoring of the functional status of older adults living in the community. Initial data indicate that, although this system is a viable mechanism for collecting information about a client's status, it cannot completely substitute for a case manager's in-home assessment. Current problems with the system include limitations with respect to handling nonstandard responses (Mahoney, Tennstedt, Friedman, & Heeren, 1999).

Telephone-Linked Care for Alzheimer's Disease (TLC-AD) is an automated telecommunications system for AD caregivers currently being evaluated at the Boston site of the Resources for Enhancing Alzheimer's Caregiver Health (REACH) program. A multisite program, REACH is funded by the National Institute on Aging and the National Institute of Nursing Research, which is evaluating the efficacy of a variety of interventions with respect to reducing burden and enhancing the quality of life experienced by family caregivers' of Alzheimer's patients. The interventions are broad-based and multifaceted, consisting of psychosocial and education services, behavior interventions, environment modifications, and technology interventions such as TLC-AD, which monitors the primary caregiver's stress and health status. It also provides a voicemail caregiver support network, an "ask the expert" call option, and a respite function to provide caregivers with a break from their caregiving duties. Lund, Hill, Caserta, and Wright (1995) evaluated a series of videotapes (Video Respite) specifically designed for patients with dementia as a mechanism to provide respite time for caregivers. The results indicated that the videotapes were helpful in providing relief for caregivers. They also seemed to reduce behavioral agitation in patients.

Telecommunications technology, including the telephone and videoconferencing, also is being used for counseling or psychotherapy. Telephone-based therapy offers several advantages over face-to-face therapy. For example, it eliminates the need for both clients and therapists to travel and allows for more flexible appointment scheduling. Telephone therapy also can be less threatening to clients because it "declinicalizes" interactions between therapists and clients. Caregiver Interventions via Telecommunications (CIT) is a telecommunication system designed to deliver psychotherapy to family caregivers of elders with dementia. This system has been implemented with a sample of caregivers and currently is being evaluated. Preliminary data suggest that the system is well received by caregivers, and that they enjoy participating in the therapy sessions. However, the overall efficacy of this type of therapy as a primary intervention method with caregivers has not yet been documented (Wright, Bennet, & Gramling, 1998).

The Miami site of the REACH program also is evaluating family therapy intervention augmented by a computer–telephone integrated system (CTIS) for family caregivers. The intent of CTIS is to enhance a family therapy intervention by facilitating the caregivers' ability to access formal and informal support services. The system involves the use of screen phones (Fig. 9.4) that allow both text and voice to be transmitted during an interactive session. It is menu-driven, system and the menus are customized for each caregiver. Given that the population involves both white American and Cuban American caregivers, the system also provides both Spanish and English text and voice messages. The menus have a hierarchical structure, and the user is guided through the menus with visual and voice prompts. Features are accessed by pressing the appropriate number on the keypad.

Caregivers can use CTIS to communicate with therapists, family, and friends; to participate in online support groups; to send and receive messages; and to access information databases such as the Alzheimer's Association Resource Guide. A respite

FIG. 9.4.   Example of a screen phone.

function also is provided. In addition, CTIS provides the therapist with enhanced access to both the caregivers and their family members. For example, the system allows family members who are unable to attend therapy session because of logistic constraints to participate in therapy online. Table 9.2 presents a list of the currently available features.

To date, experience with CTIS has been very positive, with high acceptance of the system by caregivers. Preliminary data indicate that most caregivers like the system, finding it valuable and easy to use. The most common reason why caregivers use CTIS is to communicate with other family members, especially those who do not live nearby. The data also indicate that the system facilitates communication with other caregivers. Most caregivers have reported that they found the participation in the online discussion groups to be very valuable. Many times caregivers are unable to attend community support groups because of logistic problems such as mobility restrictions or difficulty arranging alternative help. In fact, anecdotal evidence has indicated that several caregivers who could not participate in

TABLE 9.2

Features Available on the (CTIS) System

Feature	Description
Place a call	The user can use the system to place a telephone call to a family or friend. In the family screen, the caregivers can spontaneously launch a conference call with up to six other family members. The order and number of linkages are determined "on the fly" by the caregiver.
REACH discussion groups	These online telephone conferences with other caregivers are held monthly. The caregivers are invited to participate in a group, and the trained professionals facilitate the groups.
Voice messaging	All caregivers and their family members are provided with a voice mail box. They may leave messages for each other and/or their assigned therapist. Both single and broadcast messaging is available.
Reminders	Therapists are able to leave reminders to caregivers (e.g., about upcoming appointments). The reminders can be made on a one time or daily basis.
Caregiver resources	The contents of the local Alzheimer's Association Resource Guide was placed on the system with active linking to the provider's telephone number, if selected by the caregiver. The services were arranged geographically to help the caregivers with their selections.
Caregiver Respite Functions	To provide some respite to the caregiver, family members were asked to develop vignettes for the patient. These vignettes typically are reminisces about pleasant past events shared by the patient. The patient is able to listen to these messages on the screen phone via the speaker or the handset.

face-to-face support groups because of physical limitations were able to participate in discussion groups using the system. Finally, caregivers have indicated that the online resource guide was useful (Czaja & Rubert, in press).

Even so, it was a challenge to get the family therapists to integrate the technology into their therapeutic practices. Initially, many were uncomfortable with the technology and reluctant to use it. However, once they understood the potential value of the system with respect to facilitating family therapy (e.g., allowing long-distance family members to participate in sessions and reducing the need for travel to a client's home), they adopted the technology more readily. In fact, many of the therapists reported that they found online therapy to be very effective. This finding suggests that in designing these types of systems, it is important to consider the needs and preferences of both health care practitioners and caregivers. Also, strategies need to be developed for system implementation that foster use of the system among both user groups. To design systems that are useful to and usable by intended populations, it is important to conduct a formal analysis of user preferences and the types of tasks they need to perform (Gould, 1997). In this case, for example, the system would have been improved by adding additional features to the menu such as information about health benefits or other services such as legal aid or community events (e.g., upcoming meetings of the Alzheimer's Association).

Future innovations in telephone and cable television networks, such as expansion in voice and video, also may prove beneficial to caregivers. For example, telemedicine is an emerging technology that may be effective in terms of improving quality of life for both caregivers and AD patients. Telemedicine may foster links among clinicians, caregivers, and care recipients so that clinicians are better able to monitor a patient's status/behavior and provide information, advice, and support to caregivers. (See Whitten, chap. 7, this volume, for a more complete discussion of telemedicine.) The benefits of this technology include increased speed of diagnosis and treatment, patient access to physician and specialty care, continuity of care, greater patient involvement in the care process, and enhanced patient knowledge and compliance (Brennan, Moore, & Smyth, 1995). Use of such technology also may result in cost savings because there is less need for a health care provider to travel to a patient's home or, conversely, for a patient to travel to the care provider's office for minor complaints. Technology of this type also may reduce unnecessary trips to emergency rooms, and ultimately may lower probabilities of readmission or institutionalization resulting from earlier diagnosis and treatment. Friedman, Stollerman, Mahoney, and Rozenblyum (1997) have designed a telecommunications system linking voice and database components that can be used as an alternative or supplement to office visits for ambulatory care. Preliminary evaluation of the system indicates that it is well accepted by patients and health care professionals and effective with respect to clinical outcomes (e.g., medication adherence).

For the benefits of technology to be realized by caregivers, patients, and health care providers, it is important that the technology be useful and usable by these populations. It also is important that the systems be reliable and responsive. If a system is unavailable, it cannot be used, and if it is unreliable, users will become frustrated and avoid using the technology regardless of its potential benefits. For example, in the study of e-mail by Czaja et al. (1993), one of the reasons why the use of the system declined over time was that the system broke down and was unavailable. The next section highlights some needed research to ensure that these basic design criteria are achieved.

FUTURE RESEARCH DIRECTIONS

Whereas information technologies hold the promise of improving the quality of life for caregivers and care recipients, a number of issues must be addressed before the full benefits of these technologies can be realized for these populations. In this regard, research in the following areas is warranted:

- A number of issues associated with system design need to be considered. For example, the relative advantages of more advanced technologies such as interactive voice and video systems compared with simpler technologies (e.g., telephones) needs to be understood. Currently, videoconferencing technology is relatively expensive and not easily accessible. Usability issues associated with the actual design of the interface, such as input device design or the structure of online aiding systems, also need to be addressed. Although Internet use is causing a user increase, those who have limited experience with technology find the Internet overly complex and difficult to use. Usability issues are particularly important for caregivers because they are faced already with many demands and do not need to be burdened with complex or cumbersome technologies.

- Optimal strategies for combining technology with other types of interventions need to be identified. For example, it is important to determine the optimal number of face-to-face versus technology-based therapy sessions, or the extent to which technology-based assessments can be replaced by home-based assessments.

- Researchers should more thoroughly document the efficacy of technology-based interventions. For example, questions arise regarding the relative effectiveness of online versus face-to-face support groups, or online versus face-to-face interaction with clinicians. Methods for documenting outcomes associated with technology-based interactions also need to be identified.

- Data also are needed regarding advantages and disadvantages of technology for health care providers. For example, it is speculated that benefits of technology for clinicians include reduced travel time and more flexible scheduling. However, these benefits may be outweighed by costs associated with the need to respond to frequent e-mail queries or inability to interact with patients in the context of their home. Policies need to be developed regarding mechanisms of compensation for physicians who offer advice via e-mail. Research also is needed to identify strategies for training caregivers and health care providers to use information technologies. Currently, little is known about teaching novices or older people how to use the Internet for effective information seeking (Czaja & Lee, 1999).

- A more formal analysis of functions useful to caregivers, patients, and health care providers needs to be conducted. The research of Czaja et al., 1993 has shown that people are much more willing to use technology if they perceive it to be useful. As discussed by Beith (chap. 2, this volume), a careful needs assessment should be conducted before the design and implementation of a technical system.

- Data also are needed on the cost effectiveness of technological interventions. This type of data is needed, for example, to help establish reimbursement policy schedules among health insurance providers. This is true for both caregivers in terms of purchasing technology and clinicians in terms of billing for patient services.

- A number of issues concerned with privacy, safety, and quality control need to be addressed. For example, many caregivers and clinicians may be afraid to discuss sensitive issues over the telephone for fear of losing privacy. These also may be issues of liability associated with automated assessment or monitoring of a patient's status. Finally, there are no mechanisms in place to monitor the quality of information people can access online. This may be particularly problematic for people such as caregivers, who may have limited time or resources for evaluating the quality of information sources.

In summary, research that promotes a better understanding of "what works" in terms of making information technology more useful, usable, and accessible for caregivers, care recipients, and health care providers is needed and timely. Technology is developing rapidly and being used in a wide variety of domains for a wide variety of applications. At the same time, family caregiving is a burgeoning social and clinical issue. Technological innovations hold promise for enhancing the ability of caregivers to provide care to family members. However, as discussed in this chapter, a number of issues must be addressed before developments in technology are truly beneficial for this population.

REFERENCES

Brennan, P. F., Moore, S. M., & Smyth, K. A. (1995). The effects of a special computer network on caregivers of persons with Alzheimer's disease. *Nursing Research, 44*, 166–172.

Czaja, S. J., Eisdorfer, C., & Schulz, R. (2000). Future directions in caregiving: Implications for intervention research. In R. Schulz (Ed.), *Handbook on dementia caregiving: Evidence-based interventions for family caregivers*, (pp. 283–319). New York: Springer.

Czaja, S. J., Guerrier, J. H., Nair, S. N., & Landauer, T. K. (1993). Computer communication as an aid to independence for older adults. *Behavior and Information Technology, 12,* 197–207.

Czaja, S. J., & Lee, C. C. (1999). *Training and the Use of the internet in older adults.* Paper presented at the German-American Academic Council Conference, Ann Arbor, May 23–25.

Czaja, S. J., & Rubert, M. (in press). Telecommunications technology as an aid to family caregivers of persons with dementia. Psychosomatic Medicine.

Family Caregiving Alliance. (1997). *Annual report: California's caregiver resource center system fiscal year 1996–1997.* San Francisco, CA: Author.

Friedman, R. H., Stollerman, J. E., Mahoney, D. M., & Rozenblyum, L. (1997). The virtual visit: Using telecommunications technology to take care of patients. *Journal of American Medical Informatics Association, 4,* 413–425.

Gallienne, R. L., Moore, S. M., & Brennan, P. F. (1993). Alzheimer's caregivers: Psychosocial support via computer networks. *Journal of Gerontological Nursing. 12,* 1–22.

Gould, J. D., Boies, S. J., & Ukelson, J. (1997). How to design usable systems. In M. G. Helander, T. K. Landauer, P. V. Prabhu, *Handbook of human-computer interaction* (pp. 231–254). Amsterdam: Elsevier.

Kennet, J., Burgio, L., & Schulz, R. (2000). Interventions for in-home caregivers: A review of research 1990 to present. In R. Schulz (Ed.), *Handbook on Dementia Caregiving,* (pp. 61–126), New York: Springer.

Leirer, V. O., Morrow, D. G.. Tanke, E. D., & Pariante, G. M. (1991). Elders nonadherence: Its assessment and medication reminding by voice mail. *The Gerontologist. 31,* (5) 14–520.

Lund, D. A., Hill, R. D., Caserta, M. S., & Wright, S. D. (1995). Video respite: An innovative resource for family, professional caregivers, and persons with dementia. *The Gerontologist. 35,* 683–687.

Mahoney, D., Tennstedt, S., Friedman, R., & Heeren T. (1999). An automated telephone system for monitoring the functional status of community residing elders. *The Gerontologist, 39,* 229–234.

Marchionini, G. *(1995). Information seeking in electronic environments.* New York: Cambridge University Press.

McPhee, L. (1998). Number of Internet users and shoppers surges in United States and Canada. *CommerceNet.* Available: *http:1*/www.commerce.net/news/press/ 1 9980824b.html.* Accessed: 8/24/1998

Ory, M. G., Yee, J. L., Tennstedt, S. L., & Schulz, R. (2000). The extent and impact of dementia care: Unique challenges experienced by family caregivers. In R. Schulz (Ed.), *Handbook on dementia caregiving: Evidence-based interventions for family caregivers* (pp. 1–33). New York: Springer.

Otten, A. (1991, April 22). About 15% of U.S. adults care for ill relatives. *Wall Street Journal,* p. B 1.

Paveza, G. J., Mensah, E., Cohen, D., Williams, S., & Jankowski, L. (1998). Costs of community-based long-term care services to the cognitively impaired aged. *Journal of Mental Health and Aging, 4,* 64–82.

Schulz, R., & Beach, S. (1999). Caregiving as a risk factor for mortality. The caregiver health effects study. *Journal of the American Medical Association, 282,* 2215–2219.

Schulz, R., Gallagher-Thompson, Haley, W., & Czaja, S. J. (2000). Understanding the intervention process: A theoretical/conceptual framework for intervention approaches to caregiving. In R. Schulz (Ed.), *Handbook of dementia caregiving: Evidence-based interventions for family caregivers* (pp. 33–60). New York: Springer.

Schulz, R., O'Brien, A., Bookwala, T., & Fleissner, K. (1995). Psychiatric and physical mobility effects of dementia caregiving: Prevalence, correlates and causes. *The Gerontologist, 35,* 771–791.

Smyth, K. A., & Harris, P. B. (1993). Using telecomputing to provide information and support to caregivers of persons with dementia. *The Gerontologist, 33,* 123–127.

Teel, D. S. (1999). Technology & senior citizen [online]. Available: *http://www.sunlining.com/seniortech.html* Accessed: April 1999.

U.S. Bureau of Census (1997). U.S. Government Printing Office, Washington, DC.

Wright, L. K., Bennet, G., & Gramling, L. (1998). Telecommunication interventions for caregivers of elders with dementia. *Advances in Nursing Science, 20,* 76–88

Wright, S. D., Lund, D. A., Pett, M. S., & Caserta, M. S. (1987). The assessment of support group experiences by caregivers of dementia patients. *Clinical Gerontologist, 6,* 35–59.

PART IV

TECHNOLOGY AND MEDICINE

10

A Patient-Centered Approach to Automated Telephone Health Communication for Older Adults

Daniel G. Morrow
University of New Hampshire

Von O. Leirer
Decision Systems

Changes in health care delivery systems and other factors have led to a revolution in health care. Patients now are viewed as information consumers and active partners in making health care decisions rather than passive recipients of services. This revolution is fueled in part by advances in technology that facilitate the distribution of health information and services to patients. For example, the federal government has called for the distribution of computer-generated information about new prescribed medications to 95% of all pharmacy patients by the year 2006 (Department of Health and Human Services, 1996). More generally, a variety of computer-mediated interactive health communication systems now provide patients with an array of services and information (Eng, Gustafson, Henderson, Jimison, & Patrick, 1999).

The research reported in this chapter focused on automated telephone messaging, an increasingly common form of interactive health communication. Automated telephone calls are used routinely by health organizations to remind patients about appointments, medications, and other services (Leirer, Morrow, Tanke, & Pariante, 1991; Tanke & Leirer, 1994; Tanke, Martinez, & Leirer, 1997). Telephone reminders have been shown to improve appointment attendance (Macharia, Leon, Rowe, Stephenson, & Haynes, 1992; Roter et al., 1998), and automated reminders make this intervention more cost effective (Leirer, Tanke, & Morrow, 1993). Automated messaging also is used to monitor community-dwelling elderly (Mahoney, Tennstedt, Friedman, & Hereen, 1999) and chronically ill patients (Patel & Babbs, 1992; Piette, 1997; Piette, Mah, McPhee, Kraemer, & Weinberger, 1999), and to provide support for long-term caregivers (Czaja, chap. 9, this volume). This technology also may improve patient education. Age-related barriers to patient education include older patients' limited literacy and cognitive abilities as well as inadequate health care provider communication (Brown, 1990; Piette, 1997; Wasson et al. 1999). Automated messaging may address these barriers by providing a cost-effective tool for tailoring health information about self-care and adherence. Because of these benefits, automated messaging plays an ever-increasing role in community-based health care (Piette, 1997) and telemedicine (Bashur, Sanders, & Shannon, 1997).

In summary, evidence suggests that automated messaging has the potential to inform, to support decision making, and to promote preventive and self-care health practices. However, this potential will be realized only if patients find the technology easy to understand and use. It is especially important that older adults benefit from this technology because they constitute a growing segment of health consumers. The proportion of older adults in the population of the United States and other countries not only is increasing, but older adults also require proportionally more health services and thus are more reliant on health communication (U.S. Department of Health and Human Services, 1990). At the same time, they are interested in using computer technology to receive information about health care and other services (Czaja, Guerrier, Nair, & Landauer, 1993). As a consequence, older adults are likely to interact increasingly with automated messaging systems. Older adults stand to reap substantial benefits from health care technology designed with the needs, interests, and abilities of older adults in mind (Czaja, 1997; Rogers, Meyer, Walker, & Fisk, 1998).

HUMAN FACTORS ISSUES FOR AUTOMATED TELEPHONE MESSAGING

Many age-related human factors issues are related to the design of automated telephone messaging systems (for reviews, see Marics & Englebeck, 1997; Schumacher, Hardinski, & Schwartz, 1995). These include hardware design (e.g., speakerphone vs. handset, size and arrangement of buttons on touch pad), voice interface design (e.g., menu breadth vs. depth), language and message characteristics (e.g., vocabulary, speech rate, message organization), and dialog structure. The current research focused on the design of the messages delivered by these systems because these messages are the most direct link between patient and health organization. Whereas previous research has shown that patient adherence and monitoring can be improved by automated messaging (Leirer et al., 1993; Piette, 1997), these studies did not focus on the messages themselves. Automated messages should be most effective if they are accurate, easy to understand, and relevant to the patient's health care goals.

Message design is especially important for older adults because they tend to experience declines in working memory capacity and processing speed (Park et al., 1996; Salthouse, 1991), which predict differences in recall of printed (Hartley, 1993) and spoken text (Stine & Wingfield, 1990). Age-related declines also occur for recall of medication information (Diehl, Willis, & Schaie, 1995; Ley, 1988; Morrell, Park, & Poon, 1989; Morrow & Leirer, 1999). Furthermore, age-related differences in literacy impair comprehension of health information (Qualls, Harris & Rogers, chap. 4, this volume). Recent research found that more than one third of community-dwelling Medicare enrollees in several samples did not understand basic information related to taking medication, operating medical devices, and attending appointments (Gazmararian et al., 1999). Therefore, health care messages must be designed with the needs and abilities of older adults in mind to help them understand and remember how to perform health activities.

AGE-SENSITIVE DESIGN OF AUTOMATED TELEPHONE MESSAGES

This discussion takes a patient-centered approach to message design. This approach requires consideration of how automated messages help older adults accomplish their health goals. Next typical health activities relevant to messaging (making appointments, taking medication) are identified, as well as processes (comprehension, memory) and knowledge that patients require to accomplish these activities. This provides a framework for reviewing research on designing automated messages that help older patients accomplish their health goals.

Cycle of Health Care Activities

Health care generally involves a cycle of patient activities, and each activity depends on several forms of communication between caregivers and patients.

Preappointment Activities.   Patients or providers first identify a need for preventive or treatment services. This may involve gathering information about illness and treatments, talking to health care professionals, and making appointments.

Appointment Activities.   This stage involves attending the appointment and consulting with physicians, nurses, and other professionals. Several types of communication are involved, including appointment reminder messages and provider–patient discussion. The managed care environment places increasing responsibility on patients to decide on their health care, thereby increasing the need for information gathering, analysis, and decision making.

Postappointment Activities.   Patients often perform health-related activities at home after the health care visit, such as taking medication, following self-care recommendations, and using medical devices. For chronic illness, this stage also involves monitoring the progress of the illness. Such disease management activities have become more critical as health care has shifted from health care facilities to the community (Piette, 1997). Postappointment communication involves consulting with physicians and pharmacists about medication, following instructions for using medication or devices, and communicating via telephone with providers about medications, illness symptoms, or other issues.

Communication at each step of this cycle has become increasingly automated, with appointment-making, appointment-reminding, medication-reminding, and disease-management services delivered by automated telephone messaging systems (Leirer, et al., 1993; Patel & Babbs, 1992; Piette, 1997). Designing automated messages that help older adults accomplish health care goals requires identifying the processes involved in performing these activities.

Processes in Health Care Activities

Health care activities such as attending appointments and taking medication require at least the following processes or components.

Attention.   Patients first seek, notice, and attend to information necessary for accomplishing the activity (e.g., when talking to providers or reading instructions). This first stage can fail if patients do not ask providers for information, or if they do not notice information presented in instructions or other health care messages. For older patients, this stage may be particularly problematic because providers tend to provide them with less information compared to younger patients, (Greene, Adelman, Charon, & Hoffman, 1986; Wasson, et al., 1999; but see Hall, Roter, & Katz, 1988).

Comprehension.   Information also must be understood, so that people learn how to perform the health activity. Appointment-keeping requires understanding when and where to go, how to prepare for the appointment (e.g., fasting for a cholesterol check), possible side effects of the procedure, and other information. Understanding requires more than remembering the text verbatim. The information must be interpreted in light of relevant knowledge and beliefs (e.g., about the illness, treatment) to create a situation model, which represents the described situations rather than the text itself (Kintsch, 1998). For procedural messages such as instructions or reminders, patients must create a situation model of the described task (Diehl & Mills, 1995). This often requires integrating new information in the message with prior knowledge by drawing inferences. For example, people may be told that they have a 10:00 a.m. appointment, and that they need to fast for 12 before the appointment. This requires them to infer that they must skip breakfast. Thus the situation model level of representation is essential for relating the message to the health care activity.

Comprehension processes such as word recognition and inferencing are coordinated in working memory, which often is conceptualized as a limited-capacity work space wherein information is stored while processed. Thus, working memory capacity is a fundamental constraint on comprehension and memory (Baddeley, 1986; Kintsch, 1998). Age-related declines in working memory capacity help to account for age differences in text memory (Stine, Soederberg, & Morrow, 1996). These declines also may influence health activities such as medication nonadherence (Leirer et al., 1991; Morrell, Park, Kidder, & Martin, 1997).

Knowledge, on the other hand, helps overcome working memory limits on comprehension and memory by facilitating retrieval from long-term memory (for a review, see Ericsson & Kintsch, 1995; Kintsch, 1998). Knowledge generally improves memory for both older and younger adults (for a review, see Hess, 1990). These findings are consistent with a patient-centered approach to communication in which information is presented in terms of what people already know about treatment or illness. Comprehension also involves metacognitive processes such as comprehension monitoring: If people decide they do not adequately understand or will not remember a message, they may initiate repair strategies such as asking for clarification, rereading (Zabrucky & Moore, 1994), or using memory aids such as notes (Einstein, Morris, & Smith, 1985).

Acceptance.   Information also must be accepted and adopted as relevant to the patient's health goals. People miss appointments or fail to take medication not only because they misunderstand the task, but also because they decide that the illness is not serious, conclude that the side effects outweigh the benefits, or respond to other motivational factors (Deyo & Inui, 1980). The impact of beliefs on acceptance and adherence is especially important for patients with chronic illness, who are more likely to have beliefs that support or undermine adherence (Janz & Becker, 1984).

Adherence Plan.   If information is both understood and accepted, patients are likely to create a specific plan for accomplishing the task. Attending appointments requires integrating task-specific information with logistic constraints (schedule, transportation). Taking medication may require integrating dose, time, and warning information about multiple medications into a complex plan. Adherence plans must be integrated with other daily plans. Problems related to planning are suggested by the finding that nonadherence increases with more complex medication schedules (Park & Jones, 1997) and for busier patients (Morrow, Carver, Leirer, Tanke, & McNally, 1999; Park et al., 1999).

Prospective Memory.   Patients must remember actually to do the task by retrieving the intended actions at the correct time or in response to cues. This type of memory is referred to as prospective memory (Einstein & McDaniel, 1996). Appointment (Oppenheim, Bergman, & English, 1979) and medication nonadherence (Park & Jones, 1997) often is attributed to forgetting.

Automated Message Design

The goal of our research is to design automated messages that improve comprehension, acceptance, planning, and prospective memory components of health activities. Such messages may reduce differences associated with age and cognitive ability by providing environmental support for performing these activities (Craik & Jennings, 1992). These messages must have appropriate content, language, organization, and presentation medium.

Message Content.   According to a patient-centered approach, health care messages should contain the information that patients want and need to perform health activities successfully (Ornstein et al., 1993). Such messages should improve comprehension by enabling patients to create an accurate situation model of the activity. These messages also may improve acceptance of the task by addressing belief-based barriers to performing the activity (Carter, Beach, & Innui, 1986; Okun & Rice, 1997). An important issue related to message content is how much information to present. Messages should be complete, but providing too much information may impair comprehension, especially for older adults. In addition, more information does not necessarily translate into better decision making by older patients (Kirlik & Strauss, chap. 5, this volume).

Language.   Automated messages should be expressed by explicit and simple language. Linguistic factors such as word length, word frequency, and sentence length have been quantified by readability formulas used to identify the reading grade level of health care messages (Morrow, Leirer, & Sheikh, 1988).

Organization.   Improving the readability of health care messages does not always translate into better comprehension, which may reflect the fact that readability formulas do not capture factors such as organization (Kintsch, 1998) and the relation of the text to reader knowledge and goals (Morrow et al., 1988; Reid, Kardash, Robinson, & Scholes, 1994). Improving organization (e.g., explicitly categorizing and presenting important information first) increases recall of doctors' recommendations (Ley, 1988) and medication instructions (Morrow & Leirer, 1999; Morrell et al, 1989). The present research focused on patient-centered message organization, assuming that with experience, patients develop general knowledge structures about health care activities such as taking medication, just as they do for other common activities such as eating in restaurants. These knowledge structures in long-term memory are referred to as schemas (Hess, 1990). The authors investigated whether health care messages are remembered better when organized in terms of schemas about health care activities.

Presentation Medium.   Speech often is more effective than print for short messages, for listeners with a wide range of verbal skills, and when persuasion is important (for a review, see Morrow 1997). Therefore, speech may be an ideal modality for some types of health care messages. The impact of interactive speech characteristics of automated messages on comprehension was investigated in this research, with a focus on message repetition. The benefits of organization for printed as well as spoken messages were compared to test whether some message designs are better suited to print than speech.

A RESEARCH PROGRAM FOR INVESTIGATING AUTOMATED HEALTH CARE MESSAGES

First, a series of laboratory studies was conducted to investigate the impact of different message designs on older adults' comprehension and memory of information they need to accomplish health care activities. The goal was to refine messages successively at the levels of content, language, organization, and presentation medium. These well-designed messages should help older adults to create a situation model of the health care task, which could improve acceptance and planning as well as comprehension of the task. To increase the generality of the findings, messages were compared for two health activities: appointment-keeping and medication-taking.

The second phase of the project involves testing whether messages that are easy to understand and remember in the laboratory also help older adults perform the health activities. The authors currently are investigating the impact of automated appointment messages on performance in a simulated appointment-keeping task. They also plan to test the effectiveness of these messages for actual patients. Because the second phase involves assessing the impact of the automated messages on adherence behaviors, it is possible to test whether these messages improve prospective memory and planning as well as comprehension and retrospective memory.

PHASE 1. AUTOMATED MESSAGES AND MEMORY FOR HEALTH INFORMATION

First, the impact of automated messaging on laboratory measures of comprehension and retrospective memory for health information was investigated. The major findings described in this section show that older and younger adults share preferences for what information should be included in automated appointment and medication messages. They also have similar preferences for how this information should be organized into messages, suggesting shared schemas related to daily health activities. Automated messages are better understood and remembered when organized to match these schemas. This finding suggests that a patient-centered approach to communication can capitalize on commonalities across patients. Comprehension is further improved when messages are repeated, although repetition is more likely to reduce differences associated with age and cognitive ability when it is imposed rather than optional. This latter finding has important implications for the design of optional features in automated systems.

Message Content

The first step was to identify content for automated health care messages. Nine possible items about appointments were identified from messages used by actual health organizations and from a review of the appointment adherence literature (Morrow, Leirer, Carver, & Tanke 1998). Long, medium, and short messages were created in order to examine in later studies how length and organization influence comprehension and memory. These versions were created on the basis of findings from a preliminary study in which younger and older adults judged whether each type of information was necessary or optional for a spoken appointment reminder message. They were told to assume that they already knew about the appointment, and that this message would just remind them to attend the appointment. Short versions contained the five items most often judged by older and younger participants as necessary for telephone reminders: clinic name, patient name, time, location, and preappointment procedure. Medium-length messages contained these "core" items as well as the appointment purpose and the person to call in case of questions, and long messages also contained items about possible side effects of the service and information intended to encourage adherence. Thus, messages were created with information that older and younger adults need to attend appointments.

The content for automated messages about taking medication also were identified (Morrow, Carver, Leirer, & Tanke, 2000). A review of existing instructions and the medication adherence literature suggested that the following items are important: medication name, doctor's name and phone number, purpose, dose, schedule (i.e., medication times), duration, warnings (e.g., do not take with alcohol), mild side-effects, severe side-effects, and emergency 911 information. Younger and older adults in a preliminary study judged whether each item was necessary or optional for a spoken reminder. Short messages contained six items that most frequently were judged as necessary: dose, schedule, duration, warnings, mild side-effects, severe side-effects. Long messages contained all 10 items. In terms of the medication-taking schema identified in the authors' earlier work on medication instructions (Morrow, Leirer, Andrassy, Tanke, & Stine-Morrow, 1996), participants thought that shorter reminders should contain information about how to take medication and possible outcomes, but not about general information.

Language Factors

Following earlier recommendations (Morrow et al., 1988), it was ensured that all messages contained simple and explicit words and sentence structure (Table 10.1). Explicitness reduces the need for inferences in understanding messages. This should benefit older adults who have particular difficulty drawing inferences from text (Diehl et al., 1995).

It also was ensured that prosodic (e.g., speech rate, intonation patterns) and linguistic characteristics of the messages would be appropriate for older adults. The messages were recorded and presented by the TeleMinder Model IV system (see a later section for more information about this system). All messages were recorded digitally by a professional male announcer (i.e., synthetic voice was not used) with normal conversational intonation and a speech rate typical of automated messages (e.g., 220 syllables per minute). The automated system records speech at the loudest level possible without introducing distortion. Telephone handset volume was adjusted individually for participants in this study to minimize hearing-related problems.

Message Organization

Whereas patient-centered approaches to health communication have focused on message content and language (Ornstein et al., 1993), this approach was extended to message organization in the current study. Because little research exists on the organization of automated messages, this study investigated whether people share preferences for organizing appointment information, which would suggest a shared

TABLE 10.1

Example of a Seven-Item, Schema-Compatible, Appointment Message

This is a prerecorded message for Terry Smith.

You have an appointment with the Seacoast Clinic.

Your appointment is scheduled for Oct. 4 at 3:00 p.m.

The purpose of the appointment is to have your blood cholesterol level checked.

Please remember to keep a record of your diet for 2 days before the appointment.

The Seacoast Clinic is located at Lyndon Hospital, 391 Brookstone Ave., on the second floor.

Please call Sue at xxx-xxxx if you have any questions.

---

schema for attending appointments (Morrow et al., 1998). Participants arranged the appointment items to create their own short, medium, and long reminder messages. Cluster analyses of the resulting orders showed that participants preferred the following organization: (a) orienting information (e.g., patient and clinic name), (b) critical task information (e.g., appointment time), and (c) pre- and postappointment information (e.g., whom to call, side effects). Organization did not vary with age, cognitive ability (but see Morrow, Leirer, Andrassy, & Tanke, 1995), or type of service described by the message (e.g., hypertension or diabetes screen).

The organization of medication messages also was investigated, following up earlier work showing that older and younger adults share preferences for organizing information into instructions for taking medication (Morrow et al., 1996). Participants used the same organization for reminder messages that they had used for the instructions (Morrow et al., 2000). In both cases, both older and younger adults organized the information into three categories: (a) general information (e.g., medication name and purpose), (b) procedural information (e.g., dose and time), and (c) possible side effects. Thus, medication schema organization generalized across communication goal (to instruct or to remind) as well as age. This age-invariance in health-related schemas helps to provide an empirical basis for a patient-centered approach to designing health-related communication. These findings also suggest that message organization can be standardized for a variety of contexts and age groups.

Schema organization also improved memory for health-related automated messages. Older and younger adults better understood and remembered schema-compatible appointment messages, whether these messages were presented visually or auditorially by an automated telephone messaging system (Morrow et al., 1998). Longer messages that included nonessential information reduced comprehension and memory, even for the "core" information, which was the same across message lengths (e.g, appointment time). This is similar to the well-known list length effect in verbal learning, in which memory for any given item declines as list length increases. Thus, adding information to automated messages can reduce memory for critical information. Similar effects of organization and length occurred for automated messages about taking medication (Morrow et al., 2000). Older and younger participants benefitted to the same degree from appointment and medication message organization.

Schema organization especially helped listeners to draw inferences from medication messages, suggesting that organization facilitates the process of creating situation models (Morrow et al., 2000). If the situation model level of comprehension supports the ability to develop adherence plans, schema-compatible message organization may facilitate planning and performance of the health activity as well as comprehension.

Message Presentation Medium

The authors investigated whether comprehension and memory is improved by tailoring message presentation to patients, focusing on repetition because it is a common feature of automated messaging systems. Repetition may reduce the impact of age-related declines in working memory and speed of processing on comprehension by providing an opportunity for review, so that listeners are able to correct or elaborate their initial interpretation. Age differences in list and text recall can be reduced when presentation is self-paced, especially when review is possible (Verhaghen, Marcoen, & Goossens, 1993). Repeating spoken messages provides opportunity for structured review.

Imposed Message Repetition. The impact of imposed repetition (message presented two to three times) on memory for appointment information was investigated (Morrow, Leirer, Carver, Tanke, & McNally, 1999a). Repetition improved memory by 15% to 20% (also see Morrow et al., 1995). Most importantly, one repetition reduced differences associated with age and cognitive ability (as measured by working memory and processing speed tasks). However, this was the case only when retrieval support was provided (in the form of questions rather than free recall), suggesting the importance of environmental support during both comprehension and retrieval for improving older adult memory (Bäckman, Mäntylä, & Herlitz, 1990). Repetition eliminated differences in answering questions about inferred as well as explicitly stated information, suggesting that it especially helped listeners create a situation model of the appointment task.

Optional Repetition. Because repetition usually is optional rather than imposed in automated messaging, it is important to identify who takes advantage of this interactive feature (Morrow, Leirer, Carver, Tanke, & McNally, 1999b). Optional repetition will improve comprehension only if listeners realize that they did not fully understand the message after one presentation, and that they would benefit from review. Optional repetition should reduce age differences if older listeners accurately monitor comprehension and then correct or elaborate their initial interpretation. To maximize memory, they should listen more times than younger adults.

On the other hand, optional repetition may be less likely than imposed repetition to reduce age differences because of increased demands on self-initiated processing. For example, listeners must decide whether to repeat the message and then perform an action to do so (press a key). This could be a disadvantage to some older adults because of declines in self-initiated processing (Craik & Jennings, 1992).

The participants in this study listened to each message as many times as they wanted, predicted their level of recall (on a 7-point scale), and then recalled the message. Both older and younger adults repeated messages equally often (usually 1 or 2 times). Repetition improved memory for both age groups, but did not reduce age differences in memory for the total message or for the core information in the message (Morrow et al., 1999b). Age declines in self-initiated processing may have interfered with older adults' ability to take full advantage of optional repetition. Next, the study investigated whether older adults would benefit more from taking notes while listening to the messages (Morrow et al., 1999b).

Optional Repetition and Note-Taking.   Older adults often report using strategies such as note-taking for everyday memory tasks (Maylor, 1996). In addition to serving as an external storage aid that may support prospective memory, taking notes may improve comprehension and memory by focusing attention or helping listeners to reorganize information (Einstein et al., 1985). Note-taking has been shown to reduce age differences in word list recall (Burack & Lachman, 1996). As in the previous study, the participants listened to each message as many times as they wanted before the memory test. They also could take notes while listening (although they could not refer to the notes during testing). It was found that most participants took notes, and that there were minimal age differences in the completeness and accuracy of these notes. Older adults were especially likely to take accurate notes from organized messages, and to include core information (e.g., appointment time) in these notes. Therefore, message organization supported note-taking. As in earlier studies (Morrow et al., 1998), organization improved and length reduced memory, although participants tended to listen to the less-organized and longer messages more times.

Both note-taking and message repetition improved older and younger adults' memory for the messages. However, age differences were not reduced by the use of these strategies. This could reflect age differences in self-initiated processing, meta-cognitive processes, or both. Because of declines in self-initiated processing, older adults may be less likely to take full advantage of repetition (for similar proposals see Burack & Lachman, 1996 for note-taking, and Murphy, Schmitt, Caruso, & Sanders, 1987 for study time). Consistent with this interpretation is the finding that participants with higher levels of cognitive ability benefitted more from optional strategies in the current study, whereas imposed repetition reduced differences associated with ability in Morrow et al. (1999a). Note-taking also may have increased demands on self-initiated processing in the current study. For example, older adults may have had difficulty coordinating note-taking and listening to the messages. Older adults' recall of spoken medication information has been found to improve with note-taking only when the information is presented slowly and with repetition ("elderspeak" presentation), but not when the information is presented with faster "normal" speech (McGuire & Morian, 1999).

Age differences in metacognitive processes also may hinder older adults' use of optional strategies. They may not realize that they need to repeat messages because of poor comprehension monitoring. However, two findings suggest that age differences in monitoring were minimal in the current study. First, older but not younger adults repeated the less organized messages more times than the schema-compatible messages, suggesting that they were sensitive to comprehension demands. Second, age differences in the accuracy of predicting how well messages would be recalled were not significant in the current study (also see Zabrucky & Moore, 1994). Another possibility is that older adults had trouble coordinating monitoring with the additional processing needed to benefit from review (Dunlosky & Connor, 1997). For example, Zabrucky and Moore (1994) found that older adults were just as likely as younger adults to notice inconsistencies when reading, but less likely to reread the prior text in order to clarify the inconsistency.

Although the focus in this study was on the finding that optional strategies do not reduce age differences in memory, it is important to point out that older adults did take advantage of these strategies. Both older and younger adults benefitted from brief, well organized automated messages that reduced the need for repetition and supported their note-taking, which can serve as a prospective memory aid that compensates for age-related memory declines. The next section describes a preliminary study of appointment attendance suggesting that automated messages also support prospective memory.

PHASE 2. AUTOMATED MESSAGES AND APPOINTMENT ATTENDANCE

To investigate whether well-designed appointment messages that are easy to understand in the laboratory also help older adults attend appointments (Morrow, Carver et al., 1999), a simulated appointment scenario was used. With this method, participants call the automated system at prearranged times instead of actually attending appointments. Five questions about the impact of automated messages on complex health-related behaviors were investigated: (a) Does the automated system collect data about appointment adherence and participants' interactions with the automated messages at home? (b) Is the automated system easy to use? (c) Do automated messages improve adherence? (d) Does aging influence adherence? (e) Do individual differences in adherence relate to patient and situational factors? The findings from this study help to develop a multifactor model of adherence and suggest recommendations for tailoring message content, organization, and presentation to patients.

Design and Predictions

The evening before each appointment, participants receive either a well-designed (7-item, explicit, schema-compatible) message, a poorly designed (7-item, noncompatible) message, or no message at all. Half of the participants in each condition were young-old (60–74 years) and half were older (74–85 years). At this writing, the data had been collected for the well-designed message and the control conditions.

Adherence measures included time deviation (call time minus appointment time, with negative values indicating early times and positive values indicating late times), late calls (percentage of calls more than 15 min late, corresponding to how long health organizations generally hold appointments), missed calls, and call completeness (how much information about the appointment participants provide when they call the system; all the participants were instructed to indicate what they would need to know actually to attend the appointment). Retrospective memory for the messages (at the end of the study) was measured. The participants also completed vocabulary, working memory capacity, and processing speed tasks; questionnaires about health-related locus of control beliefs (e.g., belief in the role of fate in one's own health) and self-efficacy beliefs related to appointment-keeping; and questions about level of activity, reminding strategies used during the study, and ease of using the automated system.

Well-designed messages should increase adherence relative to the no message group. This improvement could reflect better comprehension and retrospective memory for appointment information, better prospective memory, or better planning. Well-designed messages also should improve adherence relative to less organized messages. The current laboratory studies suggested that the well-designed messages are easier to understand and remember. Although participants listen to the poorly organized messages more times, they do not remember them as well (Morrow et al., 1999b). These messages may not increase adherence at all, as compared with the control condition, if they do not support comprehension and retrospective memory.

The oldest participants are more likely to experience cognitive declines that influence comprehension and memory (Park et al., 1996) and prospective memory (Cherry & LeCompte, 1999). Thus, they may be less accurate than the young-old participants. Such age differences have been found for medication-taking (Morrell et al., 1997; Park, Morrell, Frieske, & Kincaid, 1992).

Automated Appointment Messages

The messages in this study contained three sections: (a) The contact section helped to ensure that the intended person was reached. For example, if the call was answered by someone other than the participant, that person was instructed to get the participant or indicate that the system should call back. (b) The appointment section was the same as the messages in the earlier laboratory studies (e.g., reminders for hypertension and diabetes screens; see Morrow et al., 1998 for detail). (c) The concluding section allowed participants to repeat the appointment information (by pressing "1") or to continue to a concluding message.

The only difference between the two appointment message conditions in this study was the appointment section, with information presented according to older adults' schema or in a different order. However, the nonschematic messages were less scrambled than those in the earlier laboratory studies because all the messages began with the same contact section that identified the clinic and addressee, whereas the messages in the earlier studies had presented this information later in the message. Therefore, unlike the messages in the earlier studies, both versions of the messages in the field study were conversationally appropriate. Nevertheless, the authors have found that these less scrambled messages are not as well remembered as the schema-compatible messages, presumably because the appointment information is presented in a nonschematic order. To show this, 16 older adults listened as many times as they wanted to the two types of messages before a memory test. The schema-compatible messages were recalled more efficiently (37% vs. 32% of the items per message presentation, $t[15] = 2.5$, $p < .05$). This finding is theoretically important because it suggests that the organizational effects in the laboratory studies were not simply the result of disrupting a general conversational schema. Instead, the participants appeared to use a more specific appointment schema to understand the messages.

Automated Telephone Messaging System

The TeleMinder Model IV system was used in both the field study and the laboratory studies described earlier. This system was composed of three components: (a) a standard microcomputer, (b) specialized hardware that interfaced with the public telephone system to place calls, and (c) software that controlled the calling operations. Appointment reminder messages were sent by this system to participants at home the night before each appointment. To maximize contact, the system was programmed to leave messages in the event that an answering machine picked up, or to call back every 30 min up to five times if there was a busy signal or no answer. The automated system also collected adherence data. Participants called the system at prearranged dates and times, leaving a message about what they would need to know actually to attend the appointment (e.g., location, preappointment procedures). The system recorded the message, as well as the call date and time.

Procedure

The experimenter called the participant several times during the first 2 weeks to "make" 12 appointments. These calls were recorded so that participants' initial comprehension of the appointment information could be scored (there were no condition or age differences for this measure). The 12 appointment dates were spread over a 4-week period after the appointment-making phase of the study. To keep each appointment, participants called the automated system at the appropriate time.

Preliminary Results and Discussion

Altogether, 24 adults participated in the well-designed message condition, and 24 participated in the no-message control condition (12 young-old and 12 old-old in each condition).

Interaction With the Automated System. The automated system reliably recorded information about how the participants interacted with the automated messages at home. For example, whereas approximately half of the automated messages were left on answering machines, all the calls answered by a person reached the correct person. Findings show that 20% of these calls were repeated one or more times by the participants.

Usability of the Automated System. Older adults found the automated system easy to use. Participants in the message condition rated the reminder messages as useful (1 = not useful, 5 = very useful, mean rating, 3.9) and easy to understand (1 = easy, 5 = very difficult, mean rating, 1). They also found it easy to interact with the system by pressing numbers on the keypad (1 = easy, 5 = very difficult, mean rating, 1.1; for more information see Morrow, Carver et al., 1999). Comments from the participants emphasized the value of the messages for reinforcing or confirming their own external aids.

Automated Messaging and Adherence. Preliminary nonadherence rates were similar to estimates of actual nonadherence for high functioning adults (Table 10.2). The participants failed to call for 8% of their appointments, which is similar to the 5% to 15% rates of actual nonadherence for private family practices, which also involve high-functioning populations (Bigby, Giblin, Papius, & Goldman, 1983; Oppenheim et al., 1979).

The participants who received automated messages were better prepared for their appointments. The participants in the message condition called earlier than those in the control condition ($F[1,46] = 4.8; p < .05$). The analysis of variance (ANOVA) included the working memory and processing speed measures as covariates because control participants had higher scores on these measures (Morrow, Carver et al., 1999). Whereas the effect of messaging on missed calls was not significant, the 25% reduction in nonadherence was similar to that of other studies (Leirer, et al., 1993). There also was a tendency for participants who received messages to report more appointment information when they called $F[1,46] = 3.0; p < .10$), suggesting that they were better informed. This difference did not reflect better comprehension of information when making the appointment, and there was no evidence that these participants had superior retrospective memory for the information at the end of the study. It is possible that the automated messages improved prospective memory for critical appointment information.

Age and Adherence. The older participants called later than the young-old participants (time deviation: $F[1,46] = 9.5; p < .01$; late calls: $F[1,46] = 4.3; p < .05$). Other studies have found that medication nonadherence is higher for this age group (Morrell et al., 1997; Park et al., 1992). The age X condition interaction was not significant for the adherence measures in the current study. Therefore, the automated messages helped the oldest adults, but did not reduce the age difference. This was the case even though the oldest participants repeated messages at least as much as the younger participants did. This may reflect the fact that optional repetition was used in this study, and this form of repetition did not reduce

TABLE 10.2

Appointment Adherence for Older and Young-Old Participants
in Message and Control Conditions

	Deviation from Appointment Time (min)	Percentage of Late Calls (>15 min)	Percentage of Missed Calls	Call Completeness (Percentage of items reported)
Control Older[a]	+19.0	31.4	9.1	68.9
Young-Old[b]	+7.7	20.0	9.5	77.7
Mean	+13.4	26.0	9.3	73.3
Message Older[c]	+12.1	24.0	7.2	85.0
Young-Old[d]	-16.4	17.0	7.1	84.0
Mean	-2.1	21.0	7.1	84.4

[a]Age range, (75–86 years)
[b]Age range, (62–74 years)
[c]Age range, (76–81 years)
[d]Age range, (62–74 years)
(From Morrow, Carver et al., 1999)

age differences in memory for the messages in the authors' earlier studies. These findings underscore the importance of tailoring health communication to adults in their mid-70s or older.

Patient and Situational Factors in Adherence.   Findings also showed that individual differences in adherence were associated with patient and situational factors, as predicted by multifactor models of health behaviors (Park & Jones, 1997). Regression analyses suggested that health beliefs, but not cognitive ability, influenced missed calls, with a more externally directed locus of control associated with nonadherence (Morrow, Carver et al., 1999). Although older adults with more externally directed beliefs sometimes experience deficits in comprehension or retrospective memory (Morrow et al., 1998), there was no evidence for this in the current study. The influence of health beliefs on memory and adherence may relate to different preferences for health information and its use (for a review, see Rogers, 1997). Different health beliefs also may influence acceptance of the appointment task. Such relations have been found for a variety of behaviors related to prevention and treatment of health conditions (Janz & Becker, 1984).

Participants who reported being engaged in more hours of activity per week were more likely to make late calls. Level of daily activity also has been found to predict medication nonadherence (Morrow & Leirer, 1999; Park et al., 1999). Competing activities may interfere with the time-monitoring component of prospective memory tasks or make it more difficult to integrate adherence plans with other daily plans (Park & Mayhorn, 1996). Well-designed reminders may help mitigate the impact of hectic schedules on adherence.

Older participants were more likely to make late calls. This age difference did not reflect lower cognitive ability because the age groups did not differ in terms of working memory, speed of processing, or vocabulary measures (also see Morrell et al., 1997). There was some evidence that the older participants' performance on the appointment-keeping task was more influenced by competing activities. This was shown by an ANOVA for late calls, with age, condition, and activity level (above or below the median hours of activity per week) as factors (the older and young-old groups did not differ in total hours of activity). Busier participants tended to call later for the older group (41% vs. 24%), but not for the young-old group (19% vs 20%; age X activity level interaction: $F[1,40] = 3.7; p = .06$). It is possible that time-monitoring for the oldest participants was more susceptible to interference from competing activities, or that these oldest participants were less able to integrate the appointment adherence plan with other daily plans. However, it is unclear why this should be so, given that they did not differ from the young-old adults in cognitive ability.

These preliminary conclusions will need to be bolstered by studies with larger sample sizes. This will allow researchers to test directly whether automated message organization influences adherence, and to assess the impact of individual differences in cognitive abilities and knowledge on adherence, and whether these differences are influenced by automated communication. They also will be able to examine more directly the impact of real-world potential barriers to telephone communication, including the extent to which the effectiveness of automated messages is reduced by answering machines, by reaching the wrong person, distraction and noise in the house, and other factors. Ultimately, findings from these simulation studies must be validated by tests with actual patients and providers, which also will address potential barriers such as limited transportation and financial resources.

CONCLUSIONS

Findings from the laboratory studies and field study have implications for the design of automated health care messages as well as for theories of adherence to health care regimens.

Design of Automated Messaging Protocols

Message Design.   The current research suggests the following general recommendations for designing patient-centered automated messages about health care. The following also should apply to other types of health communication, such as provider–patient consultations.

> 1. *Use explicit language.*   Explicit messages may be most appropriate for older adults, in part because such messages reduce the need for drawing inferences.

2. *Use short messages.*   Automated messages should contain only information that is necessary to accomplish the described task. Memory for critical information (e.g., appointment time) is reduced when optional information is added to the message. This recommendation is consistent with patient education guidelines that recommend the use of simple messages containing critical procedural information (ad hoc committee on health literacy for the Council on Scientific Affairs, AMA, 1999).

3. *Use schema-compatible messages.*   The current research used a patient-centered approach to message organization as well as content and language. The finding that older and younger adults share schemas about appointment and medication information suggests that people think about common health activities in similar ways. Automated messages are better remembered when organized in terms of these schemas. These findings suggest that message designers can capitalize on commonalities in the representation of health activities across patients.

4. *Repeat messages.*   Repetition, a standard feature in automated messaging, improves message memory for older adults, and can reduce differences in memory related to age and cognitive ability. Repetition may be a powerful way to improve comprehension and memory in a variety of situations. For example, elderspeak presentation has been shown to improve the memory of older adults for medication information (Gould & Dixon, 1997; McGuire & Morian, 1999). Elderspeak is defined by several features that may influence memory, but repetition and elaboration appear to be especially beneficial, whereas other features (e.g., exaggerated intonation, simplified language) may antagonize older listeners (Kemper & Harden, 1999). However, it was found in the current study that repetition reduced age differences in memory only when it was a required rather than an optional part of message protocols. This finding raises issues concerning the use of optional features in computer systems.

Cognitive Requirements for Using Optional Features in Automated Communication.   Automated systems that provide a range of options are potentially useful to a wider range of users and are more powerful. For example, Piette (1999) found that a diverse sample of diabetic patients who responded to automated monitoring protocols also tended to listen to optional modules with information about self-care, diet, and other topics. The human factors literature on voice messaging suggests that these options must be clearly labeled and organized to facilitate navigation through the system.

The current findings suggest that optional features such as message repetition also can improve comprehension of critical health information. However, this study suggests that these features may not be fully exploited by people with lower levels of cognitive ability, perhaps because they do not realize how much they would benefit from them. This suggests the importance of emphasizing the potential benefits of using such features. It also is possible to reduce demands on self-initiated processing by making these features default rather than optional parts of protocols. Interactive protocols that provide feedback to listeners may further improve their comprehension by supporting metacognitive processes such as comprehension monitoring.

An Integrated Approach to Appointment Communication. This chapter began by pointing out that patients participate in a cycle of health care activities, from identifying a health-related problem, to attending appointments, to performing health-related activities at home. Each phase depends on several types of communication between providers and patients. Health communication research (including that of the authors) has primarily investigated components of this communication flow in isolation. However, communication throughout the health care cycle may be more effective if different types of communication are integrated. A patient-centered approach may provide the basis for doing this. For example, information presented to patients during appointment-making and reminding could be organized consistently in terms of patient schemas about appointments. Information about medication is presented at several points of the health care cycle: during physician and pharmacist consultations with patients, on printed instructions that patients take home, and possibly over the telephone if patients call their providers for clarification (e.g., about side effects). This information could be organized consistently in terms of patients' medication-taking schemas, so that patients receive information in an expected order. Benefits of repeating this information at different points in the cycle would be enhanced by the use of a consistent framework. In this way, the different forms of communication can provide mutually reinforcing support for health activities (Morrow, 1997).

Models of Adherence to Health Care Regimens

Findings from the current study of appointment adherence are consistent with multifactor models of adherence (Park & Jones, 1997). The findings show that appointment adherence depended not only on automated communication, but also on patient factors such as health beliefs and situational factors such as patients' level of daily activity. Current models of adherence emphasize the importance of cognitive ability related to comprehension and prospective memory, and of illness representations related to task acceptance for explaining individual differences in adherence (Park & Jones, 1997). The current research suggests that the role of communication and other supports for components of adherence (e.g., note-taking and other external aids) will need to be integrated into these multifactor models. For example, the findings suggest that well-designed automated messages improve adherence by supporting prospective memory for information necessary for attending appointments. Future studies should investigate the impact of automated messaging more generally on patient education, for example, by targeting patients' illness representations. It is also important to determine whether telephone communication mitigates, or only exacerbates problems related to busy lifestyles.

Future Directions

This chapter describes a project that is in transition from carefully controlled but artificial laboratory studies of automated messaging to testing of these messages under the more complex conditions of older adults' daily lives. Future research must assess how to integrate automated messaging into existing health care systems to improve health communication for older adults. Two examples illustrate important issues that remain to be addressed. First, the potential of automated messaging for improving patient education has been emphasized. Yet, the current laboratory findings clearly show that older adults recall less information from longer automated messages, even when the messages are repeated. A pressing question is how to present older adults with information they need to perform health-related activities without overwhelming them. One promising approach is to present information visually as well as auditorially to reduce comprehension demands on working memory. The advent of affordable screen telephones has made this feasible for automated messaging systems (Czaja, chap. 9, this volume). Visual presentation (e.g., graphics such as flow diagrams) also may be most appropriate for certain types of information, such as presenting options for making health care decisions.

A second issue relates to the use of automated telephone messages for monitoring chronically ill patients. Although automated monitoring has been proven to be feasible (Piette, 1997; Piette et al., 1999), many questions remain about the best way to organize and convey complex monitoring protocols for older adults. For example, questions about patients' symptoms require thinking about a variety of conjunctive and disjunctive relations, and it is unclear whether age-related differences in working memory capacity and reasoning influence how accurately older patients respond to these protocols. These few examples demonstrate the need for collaboration among experts in cognitive aging, health communication, and system design to ensure that automated communication technology is designed with older adults' strengths and limitations in mind.

ACKNOWLEDGMENT

Support for this research provided by grant R01 AG12163 from the National Institute on Aging.

REFERENCES

Ad hoc committee on health literacy for the Council on Scientific Affairs, AMA. (1999). Health literacy: Report on the Council on Scientific Affairs. JAMA, 281, 552–557.
Baddeley, A. D. (1986). Working memory. New York: Oxford University Press.
Bäckman, L., Mäntylä, T., & Herlitz, A. (1990). The optimization of episodic remembering in old age. In P. B. Baltes & M. M. Baltes (Eds.), Successful aging: Perspectives from the behavioral sciences (pp. 118–163). Cambridge, England: Cambridge University Press.
Bashur, R. L., Sanders, J. H., & Shannon, G. W. (1997). Telemedicine: Theory and practice, Springfield, IL: Charles C. Thomas.
Bigby, J., Giblin, J., Papius, E. M., & Goldman, L. (1983). Appointment reminders to reduce no-show rates: A stratified analysis of their cost-effectiveness. JAMA, 250, 1742–1745.
Brown, S. A. (1990). Studies of educational interventions and outcomes in diabetic adults: A meta-analysis revisited. Patient Education and Counseling, 16, 189–215.
Burack, O. R., & Lachman, M. E. (1996). The effects of list-making on recall in young and elderly adults. Journal of Gerontology: Psychological Sciences, 51B, 226–233.
Carter, W., Beach, L., & Inui, T. S. (1986). The flu shot study: Using multi-attribute utility theory to design a vaccination intervention. Organizational Behavior and Human Decision Processes, 38, 378–391.

Cherry, K. E., & LeCompte, D. C. (1999). Age and individual differences influence prospective memory. *Psychology and Aging, 14,* 60–76.

Craik, F. I. M., & Jennings, J. M. (1992). Human memory. In F. I. M. Craik & T. A. Salthouse (Eds.), *The handbook of aging and cognition (pp. 51–110).* Hillsdale, NJ: Lawrence Erlbaum Associates.

Czaja, S. J. (1997). Computer technology and the older adult. In M. Helander, T. K. Landuaer, & P. Prabhu (Eds.), *Handbook of human–computer interaction* (pp. 797–812). Amsterdam, The Netherlands: Elsevier Science B. V.

Czaja, S. J., Guerrier, J., Nair, S. N., & Landauer, T. K. (1993). Computer communication as an aid to independence for older adults. *Behavior and information technology, 12,* 197–207.

Department of Health and Human Services. (1996). Prescription drug information for patients: Notice of request for collaboration to develop an action plan. *Federal Register, 6,* (166), 43769.

Deyo, R. A., & Inui, T. S. (1980). Dropouts and broken appointments: A literature review and agenda for future research. *Medical Care, 18,* 1146–1157.

Diehl, M., Willis, S. L., & Schaie, W. (1995). Everyday problem solving in older adults: Observational assessment and cognitive correlates. *Psychology and Aging, 10,* 478–491.

Diehl, V. A., & Mills, C. B. (1995). The effects of interaction with the device described by procedural text on recall, true/false, and task performance. *Memory and Cognition, 23,* 675–688.

Dunlosky, J., & Connor, L. T. (1997). Age differences in the allocation of study time account for age differences in memory performance. *Memory and Cognition, 25,* 691–700.

Einstein, G. O., & McDaniel, M. A. (1996). Retrieval processes in prospective memory: Theoretical approaches and some new empirical findings. In M. Brandimonte, G. O. Einstein, & M. A. McDaniel (Eds.), *Prospective Memory,* (pp. 115–141). Mahwah, NJ: Lawrence Erlbaum Associates.

Einstein, G. O., Morris, J., & Smith, S. (1985). Note-taking, individual differences, and memory for lecture information. *Journal of Educational Psychology, 77,* 522–532.

Eng, T. R., Gustafson, D. H., Henderson, J., Jimison, H., & Patrick, K. (1999). Introduction to evaluation of interactive health communication applications. *American Journal of Preventative Medicine, 16,* 10–15.

Ericsson, K. A., & Kintsch, W. (1995). Long-term working memory. *Psychological Review, 102,* 211–245.

Gazmararian, J. A., Baker, D. W., Williams, M. V., Parker, R. M., Scott, T. L., Green, D. C., Fehrenbach, S. N., Ren, J., & Koplan, J. P., (1999). Health literacy among medicare enrollees in a managed care organization, *JAMA, 281,* 545–551.

Gould, O. N., & Dixon, R. A. (1997). Recall of medication instructions by young and elderly adult women: Is overaccommodative speech helpful? *Journal of Language and Social Psychology, 16,* 50–69.

Greene, M. G., Adelman, R., Charon, R., & Hoffman, S. (1986). Ageism in the medical encounter: An exploratory study of the doctor-elderly patient relationship, *Language and Communication, 6,* 113–124.

Hall, J. A., Roter, D., L., & Katz, N. R. (1988). Meta-analysis of correlates of provider behavior in medical encounters. *Medical Care, 26,* 657–675.

Hartley, J. (1993). Aging and prose memory: Tests of the resource-deficit hypothesis. *Psychology and Aging, 8,* 538–551.

Healthy People 2000. Office of Disease Prevention and Health Promotion, U. S. Department of Health and Human Services, Washington, DC.

Hess, T. (1990). Aging and semantic influences on memory. In T. M. Hess (Ed.), *Aging and cognition: Knowledge organization and utilization* (pp. 93–160). Amsterdam: North-Holland.

Janz, N., & Becker, M. (1984). The health belief model: A decade later. *Health Education Quarterly, 11*, 1–47.

Kemper, S., & Harden, T. (1999). Experimentally disentangling what's beneficial about elderspeak from what's not. *Psychology and Aging, 14*, 656–670.

Kintsch, W. (1998). *Comprehension: A paradigm for cognition*. New York: Cambridge University Press.

Leirer, V. O., Morrow, D. G., Tanke, E. D., & Pariante, G. M. (1991). Elder's nonadherence: Its assessment and medication reminding by voice mail. *The Gerontologist, 31*, 514–520.

Leirer, V. O., Tanke, E. D., & Morrow, D. G. (1993). Commercial cognitive/memory systems: A case study. *Applied Cognitive Psychology, 7*, 675–689.

Ley, P. (1988). *Communicating with patients*. London: Chapman and Hall.

Macharia, W. M., Leon, G., Rowe, B. H., Stephenson, B. J., & Haynes, B. (1992). An overview of interventions to improve compliance with appointment keeping for medical services. *JAMA, 267*, 1813–1817.

Mahoney, D., Tennstedt, S., Friedman, R., & Hereen, T. (1999). An automated telephone system for monitoring the functional status of community-residing elders. *Gerontologist, 39*, 229–234.

Marics, M. A., & Englebeck, G. (1997). Designing voice menu applications for telephones. in M. Helander, T. K. Landuaer, & P. Prabhu (Eds.), *Handbook of human–computer interaction* (pp. 1085–1102). Amsterdam, The Netherlands: Elsevier Science B. V.

Maylor, E. A. (1996). Does prospective memory decline with age? In M. Brandimonte, G. O. Einstein, & M. A. McDaniel (Eds.), *Prospective memory*, (pp. 173–198). Mahwah, NJ: Lawrence Erlbaum Associates.

McGuire, L. C., & Morian, A. (1999). *Improving older adults' memory for medical information over time: The efficacy of note-taking and elderspeak*. Annual Meeting of the American Psychological Association, Boston.

Morrell, R. W., Park, D. C., Kidder, D. P., & Martin, M. (1997). Adherence to antihypertensive medications across the life span. *The Gerontologist, 37*, 609–619.

Morrell, R. W., Park, D. C., & Poon, L. W. (1989). Quality of instructions on prescription drug labels: Effects on memory comprehension in young and old adults. *The Gerontologist, 29*, 345–353.

Morrow, D. G. (1997). Improving consultations between health professionals and clients: Implications for pharmacists. *International Journal of Aging and Human Development. 44*, 47–72.

Morrow, D., Carver, L., Leirer, V., Tanke, E. D., & McNally, A. D. (1999, August). *Automated telephone messaging and older adults' appointment-keeping*. American Psychological Association Meeting, Boston.

Morrow, D., Carver, L., Leirer, V. O., & Tanke, E. D. (2000). Medication schemas and memory for automated telephone messages. *Human Factors, 42*, 523–540.

Morrow, D., Leirer, V., Andrassy, J., & Tanke, E. D. (1995). Designing health service reminder messages for older adults. *Journal of Clinical Geropsychology, 1*, 293–304.

Morrow, D., Leirer, V., Carver, L. M., & Tanke, E. D. (1998). Older and younger adult memory for health appointment information: Implications for automated telephone messaging design. *Journal of Experimental Psychology: Applied. 4*, 352–374.

Morrow, D., Leirer, V., & Sheikh, J. (1988). Adherence and medication instructions: Review and recommendations. *Journal of the American Geriatric Society, 36*, 1147–1160.

Morrow, D. G., & Leirer, V. O. (1999). Designing medication instructions for older adults. In D. Park, R. Morrell, & K. Shifren (Eds.), *Processing of medical information in aging patients* (pp 249–265), Mahwah, NJ: Lawrence Erlbaum Associates

Morrow, D. G., Leirer, V. O., Andrassy, J. M., Tanke, E. D., & Stine-Morrow, E. A. L. (1996). Medication instruction design: Younger and older adult schemes for taking medication. *Human Factors, 38*, 556–573.

Morrow, D. G., Leirer, V. O., Carver, L. M., Tanke, E. D., McNally, A. D. (1999a). Repetition improves older and younger adult memory for automated appointment messages. *Human Factors. 41*, 194–204.

Morrow, D. G., Leirer, V. O., Carver, L. M., Tanke, E. D. & McNally, A. D. (1999b). Effects of aging, message repetition, and note-taking on memory for health information. *Journal of Gerontology: Psychological Sciences. 54B*, P369–P379.

Murphy, M. D., Schmitt, F. A., Caruso, M. J., & Sanders, R. E. (1987). Metamemory in older adults: The role of monitoring in serial recall. *Psychology and Aging, 2*, 331–339.

Okun, M. A., & Rice, G. E. (1997). Overcoming elders' misconceptions about accurate written medical information. *The Journal of Applied Gerontology, 16*, 51–70.

Oppenheim, G., Bergman, J., & English, E. (1979). Failed appointments: A review. *The Journal of Family Practice, 8*, 789–796.

Ornstein, S. M., Musham, C., Reid, A., Jenkins, R. G., Zemp, L. D., & Garr, D. R. (1993). Barrier to adherence to preventive services reminder letters: The patients' perspective. *The Journal of Family Practice, 36*, 195–200.

Park, D. C., Hertzog, C., Leventhal, H., Morrell, R. W., Leventhal, E., Birchmore, D., Martin, M., & Bennett, J. (1999). Medication adherence in rheumatoid arthritis patients: Older is wiser. *Journal of the American Geriatrics Society, 47*, 172–183.

Park, D. C., & Jones, T. R. (1997). Medication adherence and aging. In A. D. Fisk & W. A. Rogers (Eds.), *Handbook of human factors and the older adult* (pp. 257–287). Hillsdale, NJ: Lawrence Erlbaum Associates.

Park, D. C., & Mayhorn, C. B. (1996). Remembering to take medications: The importance of nonmemory variables. In D. Herman, M. Johnson, C. McEnvoy, & C. Hertzog (Eds.), *Research on practical aspects of memory* (Vol. 2). Hillsdale, NJ: Lawrence Erlbaum Associates.

Park, D. C., Morrell, R. W., Frieske, D., & Kincaid, D. (1992). Medication adherence behaviors in older adults: Effects of external cognitive supports. *Psychology and Aging, 7*, 252–256.

Park, D. C., Smith, A. D., Lautenschlager, G., Earles, J. L., Frieske, D., Zwahr, M., & Gaines, C. L. (1996). Mediators of long-term memory performance across the life span. *Psychology and Aging, 11*, 621–637.

Patel, U. H., & Babbs, C. F. (1992). A computer-based, automated, telephonic system to monitor patient progress in the home setting,. *Journal of Medical Systems, 16*, 101–112.

Piette, J. D. (1997). Moving diabetes management from clinic to community: Development of a prototype based on automated voice messaging. *The Diabetes Educator, 23*, 672–680.

Piette, J. D. (1999). Patient education via automated calls: A study of English and Spanish speakers with diabetes. *American Journal of Preventative Medicine, 17*, 138–141.

Piette, J. D., Mah, C. A., McPhee, S. J., Kraemer, F. B., & Weinberger, M. (1999). Use of automated telephone disease management calls in an ethnically diverse sample of low-income patients with diabetes. *Diabetes Care, 22*, 1302–1309.

Reid, J. C., Kardash, C. M., Robinson, R. D., & Scholes, R. (1994). Comprehension in patient literature: The importance of text and reader characteristics. *Health Communication, 6*, 327–335.

Rogers, W. A. (1997). Individual differences, aging, and human factors: An overview. In A. D. Fisk & W. A. Rogers (Eds.), *Handbook of human factors and the older adult* (pp. 151–170). San Diego, CA: Academic Press, Inc.

Rogers, W. A., Meyer, B., Walker, N., & Fisk, A. D. (1998). Functional limitations to daily living tasks in the aged: A focus group analysis. *Human Factors, 40*, 111–125.

Roter, D. L., Hall, J. A., Merisca, R., Nordstrom, B., Cretin, D., & Svarstatd, B. (1998). Effectiveness of interventions to improve patient compliance: A meta-analysis. *Medical Care, 36*, 1138–1161.

Salthouse, T. A. (1991). *Theoretical perspectives in cognitive aging.* Hillsdale, NJ: Lawrence Erlbaum Associates.

Schumacher, R. M., Hardzinski, M. L., & Schwartz, A. L. (1995). Increasing the usability of interactive voice response systems. *Human Factors, 37,* 251–264.

Stine, E. A. L., Soederberg, L., & Morrow, D. (1996). Language and discourse processing through adulthood. In T. Hess & F. Blanchard-Fields (Eds.), *Perspectives on cognition in adulthood and aging* (pp. 255–290). New York: McGraw-Hill.

Stine, E. A. L., & Wingfield, A. (1990). How much do working memory deficits contribute to age differences in discourse memory? *European Journal of Cognitive Psychology, 2,* 289–304.

Tanke, E. D., & Leirer, V. O. (1994). Automated telephone reminders in tuberculosis care. *Medical Care, 32,* 380–389.

Tanke, E. D., Martinez, C., & Leirer, V. O. (1997). Use of automated reminders for tuberculin skin test return. *American Journal of Preventative Medicine, 13,* 189–192.

U.S. Department of Health and Human Services. (1990). *Healthy people 2000: National health promotion and disease prevention objectives.* (DHHS Publication No. [PHS] 91–50213). Washington, DC: U.S. Government Printing Office.

Verhaeghen, P., Marcoen, A., & Goossens, L. (1993). Facts and fiction about memory aging: A quantitative integration of research findings. *Journal of Gerontology: Psychological Sciences, 48,* P157–P171.

Wasson, J. H., Stukel, T. A., Weiss, J. A., Hays, R. A., Jette, A. M., & Nelson, E. C. (1999). A randomized trial of the use of patient self-assessment data to improve community practices. *Effective Clinical Practice, 2,* 1–10.

Wingfield, A., & Stine-Morrow, E. A. L. (2000). Language and speech. In F. I. M. Craik & T. A. Salthouse (Eds.), *Handbook of aging and cognition, 2nd ed.,* (pp. 359–416). Mahwah, NJ: Lawrence Erlbaum Associates.

Zabrucky, K., & Moore, D. (1994). Contributions of working memory and evaluation and regulation of understanding to adults' recall of text. *Journal of Gerontology: Psychological Sciences, 49,* 201–212.

Zacks, R. T., & Hasher, L. (1994). Directed Ignoring: Inhibitory regulation of working memory. In D. Dagenbach & T. H. Carr (Eds.), *Inhibitory process in attention, memory, and language* (pp. 241–264). San Diego, CA: Academic Press.

11

Communication in Health Care Tactics For Older Adults: The Case of Heart Patients

Thomas B. Sheridan
Joseph F. Coughlin
Daekeun Kim
Massachusetts Institute of Technology

James M. Thompson
MIT and Massachusetts General Hospital

The world is crying for better health care, especially as a larger and larger fraction of the population is in the older age brackets. Nearly 35 million people are 65 years or older in the United States today. The number of older adults has grown by more than 10% over the past 10 years compared with the population younger than 65 years, which grew only 8% during the same period. Over the next 20 years, the number of older adults will nearly double. Moreover, those most likely to require extensive health and related services, the oldest old (85 years and older) are the fastest growing cohort (U.S. Census Update, 1999)

In addition to the older adults themselves, caregivers will need to leverage technological innovation fully to improve the quality and delivery of care. To remain competitive, health care providers will seek ways to expand their capacity to deliver quality care in an affordable manner. Informal caregivers, such as spouses and adult children, will seek ways to extend their ability to ensure the well-being of a loved one while reducing their own stress and the anxiety of uncertainty that often is part of caregiving. Currently, one in four American families provide care to an older adult. As the senior population grows, so will the number of family members seeking ways to care for a husband, wife, father, or mother while balancing career, children, and family (Coughlin, 1999; Schulz, Czaja, & Belle, chap. 8, this volume).

Communication technology has made great strides in recent years: Multiple communication satellites circling the earth allow clear-voice or high-speed digital signals (pictures, data) from anywhere to anywhere. But can modern communications contribute much to health care? After all, doctors and patients must interact, patients must get medicines, and sometimes patients must go physically to hospitals or other facilities to be treated by expensive magnetic resonance imaging ray (MRI), and dialysis machines. Improved access to health care, especially for the growing number of older adults, is a major policy challenge that has an impact on the health of individual adults as well as overall health care costs to society. Similarly, the health care industry, including health maintenance organizations (HMOs) and physician–hospital alliances, must often "compete on the basis of cost, quality, and access to care" (Perednia & Allen, 1995). Barriers to adequate health care are often financial, but not always (Whitten, chap. 7, this volume).

In many instances, the lack of transportation or difficulty in arranging for a simple trip serves as a barrier to routine and preventive care. Somehow, transportation is always essential to move patients, doctors, medicines, or equipment. Most older adults today, and those in the future, will live in the suburbs and rural areas of the country where ready public transportation or taxi service is not always available (Cobb & Coughlin, 2000). Moreover, most older adults (80%) live alone or with a spouse. Because there is a general dearth of ready options, many of these people, particularly those in rural regions, have difficulty making routine preventive health care trips (Damiano, Momany, Foster, & McLeran, 1994; Foster, Damiano, Momany, & McLeran, 1996). This is an even greater challenge for the estimated 5 million older adults living at home with physical functional restrictions on their personal mobility (Di Pollina, Gold, & Meier, 1993).

Where urgent care transportation services receive considerable resources from public, private, social, and religious organizations, health care trips that improve wellness or might enhance the monitoring of a condition often are neglected because of onerous scheduling, lack of availability, or some other real and perceived barrier (Damiano et al., 1994; Foster et al., 1996). The primary source of transportation for older adults who do not drive are adult children or friends. As many children move from where they grew up, and where their parents continue to live, this option becomes less available. Moreover, friends in the same age cohort are less available to drive because of their own diminished capacity or, for the very old, the social network is reduced because of natural attrition.

A new generation of telemedicine technology applications that leverage powerful and affordable information technologies may serve to improve access to routine care, and to increase the awareness of patients and their caregivers, both formal and informal, about wellness management and about seeking treatment in response to a health care event. In addition to ensuring that patients, caregivers, and facilities are aligned, how might these communications technologies improve the quality of care given? For example, numerous conditions or diseases, including congestive heart failure, diabetes, and asthma, require continuous attention from both the patient and caregiver. Even pharmaceutical compliance (i.e., taking medication on time in the right dosage) or regular physical care and other activities would benefit from in-teractive systems to prompt the patient and report their "vitals" or well-being to the caregiver. What are the trade-offs? What can be anticipated for the future?

The explicit focus of this discussion is on the dilemma of a person suffering from chronic heart disease, in particular the set of symptoms called congestive heart fail-ure (CHF), a condition that affects nearly 5 million people in the United States. It is the final manifestation of various primary diseases such as coronary artery disease, hypertension, valvular heart disease, genetic disorders, diabetes, and its sequel of infection or toxic exposure. In recent years, CHF has been the most common hospi-tal discharge diagnosis in persons older than 65 years of age.

Many devices based on new sensor and computer technology now are coming on the market for home use in monitoring of chronic disease, either by a home health care worker or by the patient him- or herself, and CHF is no exception. However, home-based monitoring itself is only part of the story. The diagnostic tools must be integrated into a communication network between patients and home-based health care workers and their home-based monitoring instruments on the one hand and centralized health care specialists and their computers on the other. This in-volves both new technology and new institutional arrangements.

The home-based monitoring device must have the means to measure whether a patient's condition is stable or getting worse, as calibrated to that individual pa-tient. It then must signal the patient, informal caregiver, or home health worker in a simple way to inform whether everything is all right, whether to connect auto-matically to the local hospital's computer for a more detailed analysis of the moni-tor's data, whether to call a designated specialist to examine the monitor's output and perhaps speak to the patient or home health worker, whether to urge the pa-tient to get to the hospital when convenient and indicate how soon, or whether to call an ambulance.

A system under development at MIT is described in a modicum of detail, along with the human factors aspects of using it. These include display and alerting crite-ria, considerations of human error by all of the parties involved, economic/societal factors, and some of the policy and market implications.

CARDIOVASCULAR DISEASE AND OLDER ADULTS

Cardiovascular disease (CVD) is estimated to affect 60 million of nearly 270 mil-lion Americans. Almost 1 million of these individuals die annually, which is 1 of ev-ery 2 deaths in the United States. The mortality from CVD exceeds that of all other diseases combined. An estimated five sixths of people who die from CVD are 65 years of age or older. According to the American Heart Association, the cost of CVD in 1999 was nearly $240 billion, which includes the costs for physicians and other professionals, hospital and nursing homes, medication, and lost productivity because of mortality and morbidity.

Congestive heart failure, (CHF), one CVD particularly prevalent among older adults, is estimated to affect almost 5 million people in the United States. Nearly 400,000 new cases of CHF are diagnosed, and another 1 million patients are hospi-talized annually. In 1999, the national cost of CHF alone was approximately $21 bil-lion. It is the final pathway of various primary cardiovascular disease entities such as coronary artery disease, hypertension, valvular heart disease, genetic disorders, and diabetes, among others. It also is the sequel of infection or toxin exposure.

Hospitalizations and mortality resulting from CHF have increased steadily since 1968, despite the overall improvement in mortality from cardiovascular disease. Currently, heart failure is the underlying cause of death in more than 39,000 persons annually. Listed as the first listed diagnosis for 822,000 persons in 1992, heart failure is the most common hospital discharge diagnosis in persons older than 65 years of age. The incidence of death from CHF is 1.5 times higher in black Americans than in whites. The problem will only get worse because the elderly segment of the population is increasing at a rate 5.6 times that of the other age groups. Currently, 25 million Americans are older than 65 years of age, and 2.7 million Americans older than 85 years. Over the next 50 years, the group older than 65 years will see a 140% increase as compared with 25% in the other age groups. Currently, the only cure for end-stage CHF is cardiac transplantation.

ENVISIONED MIX OF DIAGNOSTIC, COMMUNICATION, AND TRANSPORTATION TECHNOLOGY APPLIED TO CONGESTIVE HEART FAILURE

Consider the dilemma of an older person, or indeed any person, living at home and suffering from CHF. Now assume this heart patient is experiencing some new or more severe symptoms. He or she is likely to wonder "Should I wait a while? Is it just my imagination? Should I call the doctor? Should I get myself to the hospital? What should I do?" Waiting may just create more anxiety and stress. Calling the doctor usually is a less than satisfactory experience. Seldom it is possible to get through immediately. The patient must leave a message and wait for the doctor to call back. If the patient is anxious, there is the option of going directly to the hospital emergency room. But to the patient who is not feeling well or has limited energy or bodily capacity, getting to public transportation is a difficult and even frightening prospect. A taxicab may be out of the question because of cost. Waiting for a friend or family member to provide the transportation may be inconvenient or impossible. Once the patient gets to the hospital or doctor's office, a frustrating wait is likely. This is not a happy situation, yet it is so common as to be taken for granted. Need it be?

The medical and engineering communities are not unaware of this situation. There are many current efforts to design, test, and market new devices to be used in the home or nursing facility, either by patients themselves or by trained health care workers. Often, these devices are smaller and more user-friendly versions of standard in-hospital devices to measure temperature, pulse rate, blood pressure, blood oxygen saturation, and the like. Such devices also are getting smarter, with computer chips integral to the devices indicating whether the readings are in normal or abnormal range, or somewhere between. New prototypes of some such devices are getting so small and nonintrusive as to be wearable as finger rings, band-aids, or articles of jewelry or clothing. Such devices have the prospect of round-the-clock continuous monitoring of important physiologic variables.

Given these technological advances, where does this leave the anxious patient in the dilemma of what to do about the symptoms just now being experienced? It is here that the problem must be examined in a much broader context, beyond just gadget technology, to include communication, transportation, and human factors.

Assume for the moment that a variety of relatively small, nonintrusive medical monitoring devices are both available and affordable, a quite realistic prospect for 5 to 10 years hence. Further assume that the conventional telephone is augmented by a modem for data communications, nothing more than what Internet users have today. How might the way communications, transportation, and physician–hospital operations currently work together be modified?

Envision the following. The chronic heart patient has purchased or has been provided through a health maintenance organization with a heart monitoring apparatus. Some of this may be wearable and may monitor continuously, but this need not be so. In any case, it is there; it is plugged in; and it is connected (or easily connectable) to the telephone modem. In the case of older or sufficiently incapacitated patients, a loved one or a visiting health case worker can make sure that the apparatus is attached properly to the patient and working properly. For example, a self-checking circuit provides a green light that says that this is the case.

Periodically during the day or night, the apparatus automatically connects to a computer in a local hospital and provides an indication of what the data stored in the apparatus over the past so many hours imply. The apparatus does not send all of the raw data. If the data all are in the normal range, only a brief summary is sent to the hospital computer. If the data are in an abnormal range, or if the recent trend is pointing in that direction, more data are sent. In any case, an electronic record of the patient's status is updated in the hospital computer. The computer keeps track of changes, and compares the new data to what it already knows about the patient's medical history. None of this requires any breakthrough in medicine, communication, or computer science. It is essentially a matter of systems engineering—putting the technological, institutional, and human interactive pieces together so they function smoothly and reliably.

Now here is where the transportation comes in to the picture. The patient does not have to guess about whether a trip to the doctor or the emergency room of the hospital is in order. If the hospital computer (managed by the hospital or third-party health information management and call center) decides that the patient is obviously in a critically serious condition, it can alert the patient that an ambulance is on its way. If the decision is serious but not demanding of immediate attention, the computer can instruct the patient to get to the hospital as soon as convenient, requesting a telephone call back to schedule an appointment with a nurse or doctor. If the data are ambiguous but pointing toward something serious, the computer may alert a doctor or diagnostic specialist immediately to take a closer look at the data and/or telephone the patient. If the data are not normal but suggest a need for human attention soon, this can be provided on a reasonable schedule. If the data are normal the only response may be some feedback to the patient that things look alright, and that the patient should keep up the good work. In fact, some advisory feedback from the computer, a human, or both is advisable in any case.

Figure 11.1 shows the decision logic of how the proposed diagnostic system, once developed, would fit together with the aforementioned communication and transportation system logistics.

HEALTH AND ECONOMIC BENEFITS

The system just described has the potential to improve health care in several major ways. It reduces the patient's uncertainty about what to do, and hence the patient's stress level, a factor increasingly acknowledged by physicians to be a major contributor itself to making things worse. If the patient does need immediate attention, such a system reduces the time, and the variability in time, required to provide proper attention. It relieves the emergency ward of having to screen many patients who are anxious and uncertain, but who could and should be treated by their physicians at some later time or not at all. Finally, it involves the patient in taking an active and realistic interest in his or her own health, a factor not to be downplayed. Currently, dominant model of heath care delivery rests on a "mean time between failure" approach. That is, the patient generally seeks attention and information when an event or condition presents itself. The idea is to extend the time interval between "health failure" events. In-home and increasingly interactive systems, by increasing patient and family awareness and improving behaviors that promote wellness, may extend this interval.

In addition, the stress of informal caregivers, typically spouses or adult children may be reduced by managing uncertainty. Research indicates that the constant ten-

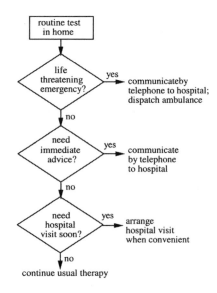

FIG. 11.1.   The decision logic.

sion and worry of informal caregivers about the well-being of a loved one is a major source of depression, illness, and lost productivity in the workplace (MetLife Mature Market Institute, 1999).

Studies also have shown that intervention can improve care and reduce costs by decreasing hospital admissions, which account for a large portion of health costs. Investigators in Los Angeles found that interventions (invasive tests, medication adjustment, patient education, and follow-up assessment at a heart failure center) decreased the number of hospital admissions from 429 in the 6 months before referral to 63 in the 6 months after referral (Fonarow et al., 1997).

The patients selected for monitoring by the intelligent system would be those at a higher risk for decompensation than the general population. These high-risk patients frequently enter the health care system too late and thus require more extensive and costly care, in addition to the emotional and physical strain they cause themselves and their families.

THE DIAGNOSTIC TECHNOLOGY

This sections and the next describe the specific technology being developed to acquire, process, and analyze blood pressure, heart rate, oxygen saturation (and possibly other signals) to make a decision on the relative health of the patient's cardiovascular system. The reader not interested in the specifics of the technology can skip these two sections.

Acquiring the Raw Data

The necessary raw medical data for the preceding variables can be acquired by existing commercial instrumentation. Blood oxygen saturation commonly is measured from a finger or ear lobe pulse oximeter, which yields a measure with every heartbeat. Similarly, the two variables of blood pressure, systolic and diastolic, are measured in a straightforward way to yield running variables. Heart rate is obtained by standard electrocardiogram (ECG) instrumentation. Figure 11.2 shows sample plots of these variables.

Perhaps more interesting for this discussion is the determination of heart rate variability, a derived measure used before by the human factors community in various contexts to indicate mental workload (Vicente, Morton, & Moray, 1987). From their study of the medical literature, the authors know that heart rate variability is a potentially important indicator of CHF.

Calculating the Power Spectrum of Heart Rate Variability

In the authors' judgment, the power spectrum of heart rate variability will provide the best means for discriminating various levels of CHF seriousness (Albrecht & Cohen, 1989; Coenen, Romeplman, & Kitney, 1977; Rompelman, Coenen, & Kitney, 1977; Rompelman, Snijders, & Spronsen,). The following analytical steps are necessary to provide the desired power spectrum. First, the ECG is acquired. Figure 11.2 shows a typical ECG signal as a function of time. The peaks are the

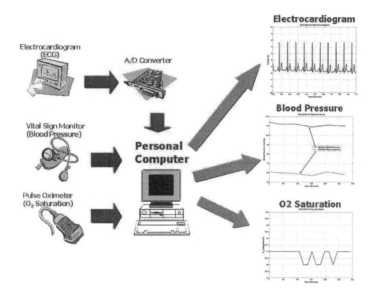

FIG. 11.2.   In-home version of conventional equipment for collecting raw data.

so-called R-wave, They are different heights, depending on the strength of the heart beat, and spaced with different time intervals between them. The problem then is to get an accurate measure of these successive time intervals. To do this, the peaks are first sharpened by a special filtering technique. The inverses of the successive time differences between R-wave peaks then are calculated (heavy vertical lines in Fig. 11.3, bottom) and held until the next R-wave peak. This provides an instantaneous heart rate. After this function has been sampled at regular intervals (thin vertical lines), a power spectrum of these samples is calculated (Fig. 11.4).

There are important details to note regarding the power spectrum. First there are roughly three peaks, at 0.005 Hz, 0.05 Hz, and 0.12 Hz. These peaks are characteristic of healthy patients in heart rate variability studies. Little energy of interest exists above 0.5 Hz.

Medical Significance of the Various Measures

The heart rate is under the control of various reflexes, the most important of which is the autonomic nervous system (ANS), which controls heart rate by two mechanisms. The first is through nerve fibers (neural response) that can either speed up (adrenergic nerve) or decrease (vagal nerve) the heart rate. The final heart rate depends on the balance of the impulses from the autonomic nervous system. Thus system also can control heart rate through a humoral response, by causing some endocrine organs to secrete hormones to change the heart rate. Other minor reflexes also control the heart rate, including atrial stretch receptors and vascular modulators. The working hypothesis is that patients with CHF have decreased vagal sig-

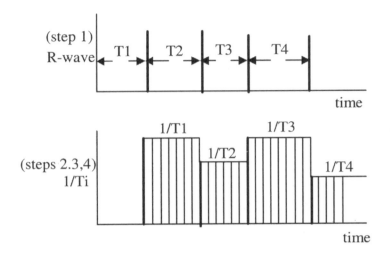

FIG. 11.3.   Derivation of heart rate variability power spectrum. (a) Measure times between R-wave peaks (upper figure); (b) take inverse of each time interval (height of bars in lower figure) and (c) hold until next R-wave peak; (d) sample result at regular intervals (thin vertical lines); and (e) compute spectral power density.

nals but unchanged adrenergic signals to the heart. Therefore, CHF can be identified by analyzing the relative amplitude of vagal and adrenergic signals to the heart (Akselrod, 1981; Berger, 1986; Fallen, 1988; Kamath, 1987).

Evaluation of the frequency characteristics of the heart rate spectrum can help to predict the relative balance between adrenergic and vagal signals to the heart. Suggestions have also been made that humoral responses also can be determined by spectral analysis. The lowest frequency peak is influenced by the peripheral vasomotor system, the muscles of the heart. The medium frequency peak is affected strongly by the baroreceptor system, and an increase power at this frequency signifies problems with electrical conduction. The highest peak is driven largely by the breathing rate.

Heart rate is important in patients with marginal coronary flow who are sensitive to the physiologic consequences of tachycardia. Tachycardia decreases coronary diastolic filling time, which decreases the supply of oxygen to myocardial tissue, especially endocardial tissue. In addition, tachycardia increases oxygen demand, which further contributes to negative myocardial oxygen balance. This results initially in regional wall motion abnormalities, which cause a rise in ventricular end diastolic and systolic pressures, further decreasing diastolic flow and starting the cycle to heart failure. Patients with mitral or aortic regurgitation can go into CHF, depending on the magnitude of the regurgitant fraction and the degree of bradycardia. Changes in the other inputs would affect the magnitude of the changes in heart rate and start the cycle toward CHF. Both an increase and a decrease in blood pressure

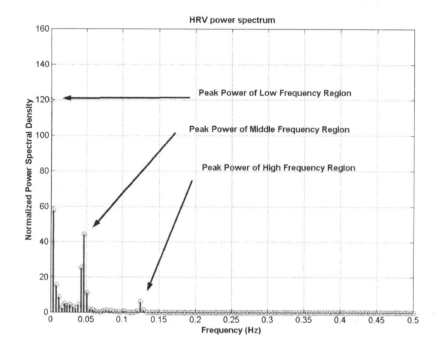

FIG. 11.4.   Heart rate variability power spectrum from experiment.

can have a deleterious effect on cardiovascular dynamics for cardiac patients. Certain types of CHF are sensitive to changes in arterial afterload, and the presence of blood pressure changes (especially systolic changes) in these patients could start the process toward CHF. Oxygen saturation, in addition to heart rate variability, is used to model quantitatively and predict the presence/direction of cardiac decompensation. Another part of the model tries to integrate other parameters such as oxygen saturation, which may indicate fluid in the lungs and failure of the alveoli to oxygenate the blood fully.

COMBINING THE DATA AND MAKING A DIAGNOSIS

A neural net (a trainable multi-input, multi-output, computer-based pattern recognizer; see Kosko, 1992) is being used as a means to combine the five signals and determine criticality of CHF. Heart rate variability and oxygen saturation are regarded as useful direct inputs to the neural net. However, well-known relations exist between blood pressure (both systolic and diastolic) and heart rate, which vary by patient age and medical history. Therefore, these three variables are normalized using some fuzzy rules before they are imputed to the neural net, as shown in Fig. 11.5.

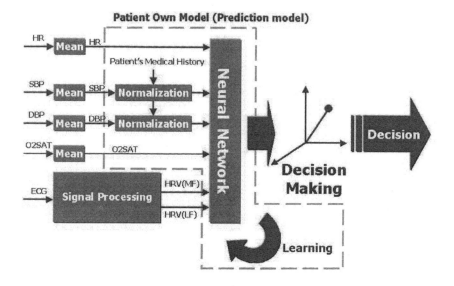

FIG. 11.5.   Use of neural net to signal level of criticality. HR, heart rate; SBP, systolic blood pressure; DBP, diastolic blood pressure; O2SAT, blood oxygen saturation; ECG, electrocardiogram; HRV, heart rate variability; MF and LF, medium and low frequency.

The output of the neural net would have several states (at least three, possibly four, to be consistent with the medical parameters discussed the preceding section on health and economic benefits. From the regions of the state space resulting thereby, a final decision is made (as illustrated in Fig. 11.1). Conceptually, the region around origin of the state space is normal. The regions successively farther out are more and more abnormal and time critical. It may be advisable not to call the ambulance until after some communication and confirmation with the patient or local health care worker.

The intention is that the neural net be trained according to Fig. 11.6. First, a model based on a range of normal healthy people would be developed. Then the several states of abnormality would trained into the neural net by comparison with patients at various stages of recovery in the hospital. The judgment of "how sick" would be made by physicians. The model for sick patients would be refined successively by prediction and feedback to determine whether it succeeded (agreed with the physicians' diagnosis) or not.

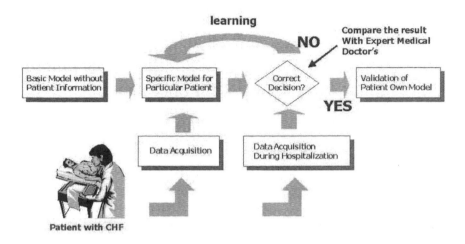

FIG. 11.6.   Procedure for using the system.

OTHER HUMAN FACTORS CONSIDERATIONS
IN THE USE OF THE SYSTEM

Providing older adults with automatic health care monitoring and computer-based advice is not immune from human factors problems. The human users, either health care workers or the patients themselves, still must understand the system enough to operate it reliably. In the following discussion, several aspects of the system design particularly critical to human use are mentioned, namely, display design, knowledge requirements and training, hardware or software failures, and human error.

Before these specifics are discussed, it should be noted that human interaction with automation is not a new problem, and there are many lessons to be learned from past experience with human interaction in aviation and air traffic control, nuclear power plants, robots, anesthesia machines, and other complex automatic systems. First, it is a mistake to think that systems are either automated or not automated. There are many levels of automation, and the level of automation can be different at different stages of a process, namely the stages of information acquisition, interpretation and display, decision making, and action implementation (Parasuraman, Sheridan, & Wickens, 2000; Sheridan, 1992). Which is best depends on many factors, at this stage of understanding, as determined empirically by simulation and field trials.

Second, an automatic system intended to relieve the human of mental workload can have just the opposite effect if the operator does not understand what the system can and cannot do, and expects it to do things it cannot do. Workload and confusion can result, especially if the system fails and the operator does not know how to cope (Hollnagel, Mancini, & Woods, 1986). The mental workload and confusion factor is particularly relevant for older adults or even for health care workers who may be nontechnical. Ideally, automation should be designed so that its capabilities and operation are easily evident from the displays, controls, and their relation to one another. Proper training also is an important requirement to enable older adults to use automated devices (Rogers, Fisk, Mead, Walker, & Cabrera 1996).

Display Design

The planned display for the in-home device is a very simple set of three lights: red, yellow, and green lights with a liquid crystal display (LCD) of text and a synthesized speech display added for redundancy. As suggested earlier, flashing red might be used to indicate "we are calling an ambulance" (emergency category of Fig. 11.1), or this might not be used if confirmation by telephone communication is deemed more desirable for avoiding false alarms signaling equipment failures and the like. Steady red means "communicate with us immediately by telephone." Yellow means "arrange an appointment as soon as convenient," and green means "continue as before." Corresponding messages would be provided in text on the LCD and by synthesized speech. This redundancy of display formats (LCD text and synthesized speech) is important for older persons whose senses in one or another mode are likely to be failing (Kline & Scialfa, 1997). By now, LCD text displays (in color) and synthesized speech are common and inexpensive.

The display for use in the hospital will involve time plots of the key variables as shown in Fig. 11.7. The proposal is that a medical specialist would monitor a display station, at which a number of patients could be checked on a time schedule according to the communicated state of criticality. A second display would enable this medical specialist to call up the medical history of any patient.

Knowledge Requirements and Training

The home health care worker is not expected to be particularly knowledgeable in CHF or even well educated. Some are professional nurses. However, many, or even most, home health workers are at a minimum level of technical expertise or education. They may be family members, or they may be employed. In any case the apparatus must be very simple to learn and to operate. With apparatus such as this, training could possibly be embedded, in the sense that the computer could guide the user through setting up and performing the tests on a step-by-step basis, providing simple troubleshooting diagnostics if the computer is not getting signals in a

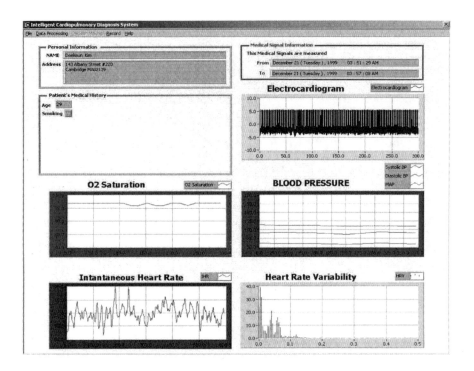

FIG. 11.7.   In-hospital display.

reasonable range (e.g., electrical connections are not properly made, electrodes are misplaced.)

Hardware or Software Failures and Human Error

The primary concern here is with errors in the home and the communications infrastructure, assuming that the hospital can afford high-reliability equipment, good maintenance, and professional personnel. However, in view of the many recent news reports of medical error in hospitals, it is clear that doctors, nurses, and hospital specialists are not immune from error.

As much as possible, the hardware and software should be self-diagnosing. That is, there should be easy means for the home operator to test the system, including the sensors, electrodes, connections, and so on. Misplaced electrodes are probably the most common human error to be expected. In this case, the system should indicate if the signals are totally out of the expected range. The National Aeronautics and Space Administration (NASA) provides some examples of techniques for checking biomedical sensor systems in-situ and alerting the users (astronauts in that case).

The thresholds at which to set the various levels of criticality is of concern? Depending on the particular patient and the way the tests are performed, there are bound to be undetected false positives (false alarms) and false negatives (serious and critical CHF symptoms) that are indicated as normal or of only mild criticality. After the equipment is made as reliable as economically feasible, the selection of thresholds is a matter of balancing the probabilities and costs of annoying false positives ("crying wolf" and the resulting loss of trust that can ensue) and the much more serious false negatives (an action after it is too late) (Green & Swets, 1960; Lee & Moray, 1992).

CONCLUSION

It is believed that the future health care system will involve far more in-home monitoring and diagnosis of patients with chronic or acute medical problems, along with moderate-bandwidth communication over conventional telephone or data channels to local hospitals. In addition, the growing availability of bandwidth via regional and national cable providers is opening up a new frontier in the range of health-related services provided to older adults in their homes. The patients themselves or home health workers, depending on circumstances, will operate the in-home data-collection equipment, and get simple straightforward advice as to what, if any, action to take. Ambulance services will be alerted automatically or manually as appropriate.

A specific system under development is described. It deals with congestive heart failure, a common disease. The instrumentation and its operation are described, along with some considerations of system validation, health and economic benefits, selection and training, and failure or error.

It is proposed that future system development follow a similar pattern complemented by an examination of how this system and others might be engineered into "networks of care" that monitor patient well-being and provide patient-specific education information to the patient, their families, and caregivers. Further development of new business, service delivery, and information management models will be conducted to assess how this and similar technologies might facilitate both wellness and care management, improving the well-being of patients and their families while reducing overall health care costs to providers and society.

ACKNOWLEDGMENTS

The authors acknowledge and thank Electronic Data Systems (EDS) Global Health Care Solutions Group for their generous support to the MIT Age Lab, which sponsored this research.

REFERENCES

Akselrod, S. (1981). Power spectrum analysis of heart rate fluctuation: A quantitative probe of beat-to-beat cardiovascular control. *Science, 213*, pp. 220–222.

Albrecht, P., & Cohen, R. (1989). Estimation of heart rate power spectrum bands from real-world data: Dealing with ectopic beats and noisy data. *Computers in Cardiology*, 311–314.

Berger, R. D. (1986). An effective algorithm for spectral analysis of heart rate variability. *IEEE Trans*. Biomedical Engineering, Vol. 9, pp. 900–904, 1986.

Cobb, R., & Coughlin, J. F. (2000). How will we get there from here? Placing transportation on the aging policy agenda. *Journal of Aging and Social Policy*, University of Massachusetts, Gerontology Institute.

Coenen, A. J. R. M., Rompelman, O., & Kitney, R. I. (1977). Measurement of heart rate variability: Part 2. Hardware digital device for the assessment of heart rate variability. *Medical and Biological Engineering and Computing*, 15, pp. 423–430.

Coughlin, J. F. (1999). Technology needs of aging boomers. In *Issues in science and technology*. National Academy of Sciences. Washington, DC: National Academy Press.

Damiano, P. C., Momany, E. T., Foster, N. S. J., & McLeran, H. T. (1994). *Transportation of rural elders and access to health care*. Ames, IA: University of Iowa, Public Policy Center.

Di Pollina, L., Gold, G., & Meier, D. E. (1993). Health care for the homebound older adult: A medical model. *The Mount Sinai Journal of Medicine*, 60(6).

Fonarow, G. C. et al. (1997). Impact of a comprehensive heart failure management program on hospital readmission and functional status of patients with advanced heart failure. *Journal of the American College of Cardiology*, 30, 725–32.

Foster, N. S. J., Damiano, P. C., Momany, E. T., & McLeran, H. T. (1996). Rural Public Transportation: Perceptions of Transit Managers, Directors of Area Agencies on Aging, and elders. *Transportation Research Record*, 1557, 58–63.

Green, D. M., & Swets, J. A. (1960). *Signal detection theory and psychophysics*. New York: Wiley.

Hollnagel, E., Mancini, G. G., & Woods, D. D. (Eds.). (1986). *Intelligent decision support in process environments*. New York: Springer-Verlag.

Kamath, M. V. (1987). Heart rate variability power spectrum as a potential noninvasive signature of cardiac regulatory system response, mechanisms and disorders. *Heart Vessels*, 3, 33–41.

Kline, D. W., & Scialfa, C. T. (1997). Sensory and perceptual functioning: Basic research and human factors implications. In A. D. Fisk & W. A. Rogers (Eds.), *Handbook of human factors and the older adult* (pp. 27–54). New York: Academic Press.

Kosko, B. (1992). *Neural networks and fuzzy systems*. NY: Prentice-Hall.

Lee, J., & Moray, N. (1992). Trust, control strategies and allocation of function in human–machine systems. *Ergonomics*, 35(10), pp. 1243–1270.

MetLife Mature Market Institute. (1999). *MetLife juggling act study: Balancing caregiving with work and the costs involved*. Mature Market Institute, Metropolitan Life Insurance Co. New York.

Parasuramsan, R., Sheridan, T. B., & Wickens, C. D. (2000). A model for type and level of human interaction with automation, *IEEE Transactions, on systems, man and cybernetics*, 30(3), (pp. 286–297).

Perednia, D. A., & Allen, A. (1995). Telemedicine technology and clinical applications. *JAMA*, 273(6), 484.

Rogers, W. A., Fisk, A. D., Mead, S. E., Walker, N., & Cabrera, E. F. (1996). Training older adults to use automatic teller machines. *Human Factors*, 38, 425-433.

Rompelman, O., Coenen, A. J. R. M., & Kitney, R. J. (1977). Measurement of heart-rate variability: Part 1. Comparative study of heart-rate variability analysis methods. *Medical and Biological Engineering and Computing*, 15, 233–239.

Rompelman, O., Snijders, J. B. I. M., & Van Spronsen, C. J. (1982). The measurement of heart rate variability spectra with the help of a personal computer. *IEEE Transaction on Biomedical Engineering, 29*(7).

Sheridan, T. B. (1992). *Telerobotics, automation and human supervisory control.* Cambridge, MA: MIT Press.

U.S. Census Update. (1999). *Older adult profile 65+.* Washington, DC: U.S. Government Printing Office.

Vicente, K. J.. Morton, D. C., & Moray, N. (1987). Spectral analysis of sinus arrhythmia: A measure of mental effort. *Human Factors, 29*(2), 171–182.

12

Designing Medical Devices
for Older Adults

Daryle Gardner-Bonneau
Bonneau and Associates
Portage, Michigan

The medical device industry traditionally has focused almost exclusively on system safety, with little regard to the usability of its products. This is not surprising given that the Food and Drug Administration's (FDA's) regulations have explicitly mandated this point of view. As a result, many medical devices are complex and extremely difficult to use, even by those trained in health care and medicine (e.g., medical technologists, nurses, clinicians). Furthermore, to the extent that the context of use was considered in the design of medical devices at all, it often was viewed as limited to the hospital environment.

With the advent of managed care, less and less care is being provided in hospitals. More and more care is being provided in the patient's home. Hospital stays are shorter than ever before. This means that medical equipment designed for hospitals and used by specialized health care personnel now is being placed in the home and used by laypersons. Users now are family caregivers or the patients themselves.

A typical scenario in home health care is the elderly woman taking care of her elderly spouse. She may have infirmities herself, probably received little or no training about the medical devices she must operate to care for her husband, and may be experiencing the stress that comes with caring for a loved one who is ill. Needless to say, medical devices were never designed with this scenario in mind. Therefore, it is not surprising that operators of complex medical equipment in the home, who may be lay caregivers or the patients themselves, often end up in the emergency room when they experience difficulties with equipment.

The good news, however, is that with the advent of the FDA's quality systems regulation (Code of Federal Regulations, 1996), medical device developers now are required to consider human factors issues in the design of their products. They must consider the capabilities and limitations of users and the context of use in the design of their devices. This chapter considers that good human engineering design can be

221

accomplished effectively in the medical device industry, which still is a cottage industry in the United States. Human factors engineering standards for medical devices (both existing devices and those under development) are discussed, along with the current challenges involved, specifically in the design of medical devices and equipment for use by older adults.

THE STATUS OF HUMAN FACTORS IN THE MEDICAL DEVICE INDUSTRY

It probably is safe to say that until approximately 5 years ago, little attention was paid to human factors in the medical device industry, with some prominent exceptions in areas such as anesthesiology and surgery. The medical device industry was shielded effectively from human factors issues in several ways. The first had to do with priorities and accountability. Medical device manufacturers answer primarily to the Food and Drug Administration and are subject to its rules and regulations (U.S. Food and Drug Administration [FDA], 1998a). In turn, the FDA's priority, with respect to medical devices, had always been safety, meaning system safety from an electromechanical perspective. Did the device or piece of medical equipment function as intended without posing a significant risk to the patient? Many safeguards existed to prevent patient injury by the action of the equipment, for example, to ensure that a machine operated correctly could not malfunction in such a way as to deliver an overdose of x-rays accidentally, or to ensure that device hardware itself was not damaged or did not malfunction because of power surges or the loss of a primary power source. The performance (including errors) of the operator, whether a highly trained professional or a lay user, was a lesser concern, and the usability of the device not a concern at all. Not surprisingly, medical equipment and device developers paid attention primarily to the FDA's requirements.

A second factor that sheltered device developers from having to consider human factors was the culture that exists in health care. Most devices were being used in hospitals by personnel subject to the "blame and train approach" that has always existed in medicine (Feinstein, 1997). It mattered not whether a device was difficult to understand, difficult to operate, or accompanied by an indecipherable instruction manual. It was up to hospital personnel to figure the device out and operate it without making mistakes. Unless a mishap was the obvious cause of a malfunctioning piece of equipment, discussions of medical mistakes and errors were buried in morbidity and mortality (M&M) conferences, at which blame was assigned and further training viewed as the means of preventing future occurrences. Despite the fact that in recent years the FDA, among others, has embraced the systems approach to analyzing medical mistakes (Berglund, 1998; Dwyer, 1997; and FDA, 1998b), this "blame and train approach" persists. In January of 1999, an article by Casarett and Helms (1999) in *Academic Medicine* demonstrated that physicians, including medical educators, have not fully accepted the implications of the systems approach. These authors wrote:

> Despite its growing following in health care, however, the systems approach creates unique risks when it is applied in an academic training program. There is reason to be

concerned that an undue focus on systemic causes of errors may be detrimental to an institution's educational mission. By exclusively focusing attention on system causes of errors, a training program may prevent its house officers from examining their own responsibility for their mistakes. They may fail to learn as much from their errors as they might. (p. 21)

It appears that these authors fail to view the physician as a component of the system, and that they still view training as the key means of error prevention. They continue to ignore Leape's (1994) admonition that systems relying on error-free performance by humans are doomed to failure.

Besides the fact that the culture of medicine, by and large, does not encourage health care professionals to take a systems view, there typically is little time for them to do so. If a medical device is difficult to use or prone to user error, the health care professional is very unlikely to have the time or make the time to follow up with a complaint or report to a device manufacturer.

The third and final factor that served to shelter medical device and equipment developers from human factors engineering concerns was the relatively limited and controlled environment in which devices were deployed (i.e., the hospital). A device developer could rely on the fact that there would be no carpeting on the floors to impede the maneuvering of devices, that electrical wiring would meet codes and requirements, that security limited the use of equipment to those with authorization, and that there would be able-bodied people available to transport and use the equipment. Many human factors considerations, therefore, were ignored by device developers because, for them, there were no consequences. It was the users in the field who had to live with or overcome the problems.

In the past 5 years, the situation with respect to the human factors engineering of medical devices and equipment has been changing, slowly but surely, from that described in the previous paragraphs. There have been at least three major drivers of this change. The first was that 1995 was a horrendous year for serious medical errors, errors so shocking that public awareness of the problem increased substantially. Ethicist Shirley Bach (2000) recently outlined the cases that occurred or came to light in 1995. In one Florida hospital, a physician operated on the wrong leg of one patient and amputated the wrong leg of another. At the same hospital, a tubal ligation was carried out without the patient's consent, and a ventilator was shut off on the wrong patient, causing that patient's death. Also in 1995, at Long Island Jewish Hospital, two anesthesia errors led to two patient deaths. In Texas, a healthy lung was removed from a patient instead of the cancerous one. A 7-year-old boy undergoing minor ear surgery in a Florida hospital received a lethal injection of adrenaline, as opposed to the lidocaine injection he was supposed to receive to stop bleeding.

Earlier, in late 1994, in the case that first widely alerted the media to the need for change, a cancer patient at Massachusetts' renowned Dana Farber Cancer Center was the victim of a series of medical errors involving a large number of people. The errors ultimately led to her receiving four times the lethal dose of her cancer medication (cytoxan). She died of heart damage, but the errors were not discovered until 3 months later when the data from the case were entered into a hospital database. Perhaps because the patient's husband was employed at the hospital, this case became the most publicized of all that occurred during this period.

A second driver for change was the FDA, which recognized that many medical errors were not the result of malfunctioning equipment, per se, but the result of user errors caused by poor user interface design (Bogner, 1993). On October 7, 1996, the FDA released its final rule (Code of Federal Regulations, Title 21, Final Rule—Quality System Regulation), which mandates in section 820.30 on Design Controls that the needs of the equipment user be considered during the design of medical devices and equipment. Specifically, it states:

> Each manufacturer shall establish and maintain procedures to ensure that the design requirements relating to a device are appropriate and address the intended use of the device, including the needs of the user and patient. (Code of Federal Regulations, April 1, 2000, p. 142).

Further, with regard to design verification and validation, the regulation states:

> Design validation shall ensure that devices conform to defined user needs and intended uses and shall include testing of production units under actual or simulated use conditions. (P. 142)

The third major driver for consideration of human factors in medical device design was the advent of managed care. Although some may credit managed care for bringing an emphasis on continuous quality improvement and error reduction to the health care arena, the impact of managed care on medical devices was, perhaps even more importantly, a result of its policies aimed at keeping patients out of the hospital as much as possible or, if they were admitted, reducing the length of stay. Care that would have been confined to hospitals was now taking place in the home. This meant that complex medical equipment would now be used in the home by people having far less training and skill than was the case in the past.

In summary, all of the factors that had isolated the medical device developer community from the reality of human factors engineering concerns are now disappearing. There is public awareness of the problem: The FDA sees human factors as a priority; the user population has changed; and the environment in which devices are used is more variable. In short, medical devices and equipment have become consumer products, and the implications of that are worrisome to device manufacturers, given the history of the industry.

MEDICAL DEVICES AND OLDER ADULTS

Older Adults as Device Users

One of the first things the author says when addressing a room full of medical device developers is "The user of your product is almost never you, and is not like you." According to Klatzky, Kober, and Mavor (1996), more than two thirds of home health caregivers are members of the patient's family. A typical caregiver is a woman older than 40 years of age who is married. More than one third of these caregivers have no more than a high school education. They are sometimes, if not often, under high levels of stress, have received limited training with the equipment they will be using, and have inadequate or no advice and support.

These new users of medical devices are unlike device users in hospitals in several other ways. First, unlike physicians, in particular, who are expected to be infallible (no matter how unrealistic that expectation is), most users of medical equipment are not subject to such professional standards of performance. They expect technology to be forgiving of error, or the consequence of error to be nil, just as it is for most other consumer products. They will abandon technology if it is cumbersome or difficult to use, sometimes even in cases wherein it may be required to save their lives (e.g., long-term oxygen therapy). Inconveniences and disruptions they would tolerate in the hospital will not be as well tolerated in the home, especially when device use is expected to be more than temporary.

The home user of medical equipment, whether the patient or not, may have age-associated disabilities or impairments that make current equipment difficult to use. Although it often is noted that the older population is heterogeneous, making generalizations difficult (Brock & Brody, 1985; Small, 1987), it nevertheless is useful to specify some of the sensory, motor, and cognitive limitations experienced by sizable percentages of this population. For example, 30% of the population between the ages of 75 and 79 years has static visual acuity of 20/50 or worse, and 9% of people between the ages of 55 and 64 years have similar deficits in acuity (Small, 1987). Similarly, the elderly as a group have a reduced ability to see in low levels of illumination, experience losses in color discrimination ability, and have a greater susceptibility to glare (COMSIS Corporation, 1988; Koncelik, 1982). Age-related hearing deficits are common, and the frequency ranges of such deficits differ between men and women (Small, 1987). Presbycusis, an age-related hearing impairment with a strong central nervous system component, the most common hearing impairment among the elderly (White, Cartwright, Cornoni-Huntley, & Brock, 1986), is associated with other signs of central nervous system impairment, including motor weakness, poor coordination, tremor, and memory loss.

Kline and Scialfa (1997) provided a recent detailed review of age-related perceptual changes. Reaction times become slower, increasing by 20% for simple reactions by the age of 60 years, with larger increases in complex reaction time. Strength also decreases significantly for many people in their 60s and declines even further thereafter. Mobility impairments also become more common with age. Studies indicate that many older people have difficulty with computer input devices requiring fine motor control and coordination. For example, Charness, Bosman, and Elliot (1995) and Casali and Chase (1993) found that many of their elderly subjects had difficulty using a computer mouse. In particular, they experienced problems with respect to clicking and dragging tasks when screen targets were small.

In addition to sensory and motor deficits, older adults may experience cognitive deficits of various types (Morrow and Leirer, 1997). Salthouse and Babcock (1991) noted that the elderly often have working memory deficits that make it difficult for them to remember complex sequences of instructions such as those commonly required for medical device operation. Older adults also are more distractable, on the average (Kane, Hasher, Stoltzfus, Zacks, & Connelly, 1994), and have difficulty both directing and inhibiting their attention in relation to irrelevant information (Hasher & Zacks, 1988; Zacks & Hasher, 1997). They also tend to be less oriented toward detail than younger people (Tun & Wingfield, 1994). McDowd, Vercruyssen, and Berrin (1991) noted that older adults experience greater difficulty than younger adults in dual-task performance situations. This is particularly true when the tasks require them to divide their attention among multiple sensory modalities (e.g., listen for auditory feedback while simultaneously monitoring a visual display, a dual-task situation quite common in the operation of medical devices; for a more detailed discussion on the characteristics of caregivers, (see Shulz, Czaja, & Belle, chap. 8, this volume.)

Home caregivers also have an emotional investment in their patients that hospital personnel do not have. They may be under considerable stress, and often cannot process equipment instructions and training that unfortunately are provided often in the hospital out of the environmental context in which that training will be used. Furthermore, home caregivers may panic in an emergency situation, are apt to forget specific, complex instructions, and may resort to more basic knowledge or instinctive behavior (Mosenkis, 1994). Such responses are by no means limited to elderly caregivers. Boodman (1999) reported that mothers at home with at-risk infants experience the same sort of stress related to complex procedures and equipment required in the home. She related the story of one young mother faced with an apnea monitor and the potential of having to remember how to administer cardiopulmonary resuscitation to her young infant in the event the baby stopped breathing. Said the mother, "I was just praying the whole time, 'Don't go off, don't go off.'"

When her baby's father told her he had to leave for work, later that day, she nearly became hysterical, saying, "Don't leave me, please don't leave."

For many users of medical devices in the home, the technology will be new and unfamiliar. They will have had little opportunity to build mental models of how devices work, nor will they be interested in doing so. In addition, some evidence exists suggesting that the building of conceptual models is not the most appropriate way to teach older adults about new technology (Caplan & Schooler, 1990). When technology appears overly complex to these users, they are apt to revert to a rote learning approach to master the operation of the device and complete their tasks (Bridger & Poluta, 1998). Whereas such an approach may serve them well under normal operational conditions, it will be problematic in novel situations or when errors occur.

The Home Health Care Environment

The home health care environment is quite a bit different from the hospital environment in which most medical equipment traditionally was operated and more highly variable. Equipment noise levels that would be tolerated in the hospital will not be tolerated in the home. Hospitals have elevators, whereas homes may have stairs and carpeting, both of which impede movement of medical devices that may be both heavy and cumbersome. Electrical wiring in some homes may be suspect, and the availability of dedicated or adequately wired electrical outlets for medical equipment may be limited. Homes also are far less secure. The presence of children, for example, in a home poses risks that equipment may be tampered with, settings changed, or machines unplugged from outlets. Such issues required only limited consideration in the design of equipment for use in hospitals, but they are critical for equipment designed to go into the home.

TYPICAL PROBLEMS IN MEDICAL DEVICE DESIGN

Both Mosenkis (1994) and Gardner-Bonneau (1997), among others, have outlined many of the typical human factors problems in medical devices. Some are general problems that would affect any user in any setting, whereas others are specific to users with various types of impairments, some of which are age related. A partial listing is discussed in the following sections.

Physical Usability

The physical operation of devices can be difficult for patients with deficits in strength, flexibility, dexterity, or vision. A good example of poor usability exists with some of the reagent strips used to monitor blood glucose. For a 70-year old diabetic who may have failing vision and limited dexterity as a result of circulation problems, the strips are small and difficult to manipulate (Laux, 1994).

Unusual or Unexpected Operation

The design of many devices fails to observe population stereotypes, and there is little standardization. Mosenkis (1994), for example, cited devices for which the "ON" switch is at the bottom of a vertical display, or on the left-hand side of a horizontal display. He also cited infusion pumps as problematic because standardization is lacking. Some pumps require the user to close a manual clamp to prevent the free flow of fluids from the infusion set. Others automatically stop the flow when the set is removed from the pump/controller. If an operator is used to one pump and encounters another, errors in operation and patient injury may result. In some cases, two conflicting population stereotypes are relevant in the same design situation. For example, gas or liquid flow typically is increased by turning a knob counterclockwise, whereas electronically controlled parameters (e.g., audio volume) usually are increased by turning a knob clockwise. With a device that controls both fluid and electronic parameters, there is bound to be confusion if both of these conventions are used.

Lack of Protective Incompatibility

Parts and accessories should connect to a device in one way, and one way only. It should not be possible to operate a device with improperly connected parts, but this problem occurs in any number of devices. For example, in blood glucose monitoring, particular reagent strips are designed for particular monitors. Strips designed for other monitors should not fit, or at least should not yield a reading if they do fit. Yet, people have been harmed when they have acted on incorrect readings, using a strip not designed for the monitor they were using (Mosenkis, 1994).

Excessive Complexity

Just like the proverbial videocassette recorder (VCR), many medical devices suffer from "requirements creep," a common problem that occurs when designers lose sight of key requirements and continuously add features or functions simply because they can, or as a result of pressure from marketing and sales divisions in the medical device company. The consequences are an excessive number of unnecessary device features and displays as well as controls that are complex and confusing. In addition, some devices provide only alphanumeric codes in the event of an equipment malfunction or error, requiring that the user read the instruction book to ascertain the cause of the malfunction. Many systems, including blood glucose monitors, also require frequent calibration, which adds to system complexity and the time required to use the device.

Defeatable or Ignorable Safety Features

Auditory alarms, in particular, are prone to being defeated (i.e., turned off or disabled), and the design of medical device alarms is one of the major challenges in an industry that lacks standardization and integration across devices. Because medical devices are designed independently by manufacturers, no consideration is given to how alarms from multiple systems or devices being used simultaneously (e.g., in an intensive care unit) will be perceived and managed by users. As a result, simultaneous auditory alarms may mask each other, or there can be situations in which the operator perceives the multiple alarms, but has no knowledge as to their priority in terms of management. In cases wherein multiple alarms are conveying redundant information, or in which the false alarm rate for a device is unacceptably high, users may choose to turn off the alarm system altogether.

Deficient Documentation and Training Materials

Documentation is neglected in many products, but poor manuals and poor training materials are especially noticeable in the medical device industry because of the complexity of medical devices and the lack of familiarity of many users with a device and its associated terminology. User manuals and instructions often are written for a professional rather than a lay user, particularly those for devices that traditionally have been used in a hospital. Couple this with the fact that what little training lay users receive often occurs in the hospital or clinic, outside the context in which they will use the equipment, and where they may be under stress and not optimally receptive to new information, and the necessity of good documentation cannot be denied.

Limited or Nonexistent Feedback About System Operation

Many systems provide little or no feedback to the user about whether the system is functioning properly or not. Feedback often is limited to responses to operator input alone. This situation does not inspire confidence, placing the user in a state of uncertainty and anxiety when operating the device or system.

Aesthetic Issues

Noisy, unsightly equipment and devices may be acceptable during a short hospital stay, but are not likely to be tolerated over the long term in the home environment. If severe enough, aesthetic considerations can result in equipment abandonment, which has been demonstrated to be a common, serious problem with rehabilitation equipment and assistive technologies as well (Gardner-Bonneau & Gosbee, 1997).

Why do these problems exist, despite the fact that enough knowledge is available to prevent many of them? Mosenkis (1994) cited the following reasons, all of which will be familiar to human factors specialists without a need for further elaboration:

The designer thinks like an engineer.

The designer is unfamiliar with the user and the use environment.

The designer is unaware that human factors expertise is available.

Designs are tested by people not representative of the user population.

The designer shifts the burden to the operator's manual.

The point is that the designers of medical systems often operate in a vacuum, failing to consider both the user and the environment of device use.

MEDICAL DEVICE DESIGN: APPLICABLE STANDARDS AND USE OF HUMAN FACTORS EXPERTISE

Much of the medical device industry is new to human factors engineering, but is paying attention now that the FDA's Quality Systems Regulation is in place. However, device developers are struggling to create processes by which human factors can be integrated into medical device design and appropriate human factors data and information can be accessed to guide design. Two approaches to meeting these goals are (a) the development of guidance and standards and (b) the direct incorporation of human factors expertise in medical device companies.

Human Factors Engineering Standards for Medical Devices

Design standards are intended to provide guidance for designers and to promote quality and consistency in key design elements (e.g., safety) throughout the industries for which they are developed. Standards represent a consensus of opinion, usually, of experts in a given field about what constitutes good design. They may include requirements as well as recommendations, but compliance with any given standard is voluntary from the standpoint of organizations such as the American National Standards Institute, one of the major standards development organizations in the United States. That said, standards can be imposed on system developers, and often are, by sponsoring organizations, in which case compliance is mandatory.

Only one human factors engineering standard in the United States is directed explicitly toward the design of medical systems and devices. That standard was published by the American National Standards Institute/Association for the Advancement of Medical Instrumentation (AAMI) in 1988 and revised in 1993 (ANSI/AAMI HE 48-1993). Unfortunately, that standard has not been very useful for a number of reasons. First, it was based largely on MIL-STD-1472D, *Human Engineering Design Criteria for Military Systems, Equipment and Facilities* (Department of Defense, 1989), and suffers from that standard's weaknesses, among them a limited coverage of user interface design for software. More important, however, is the fact that MIL-STD-1472D contains no medical examples to speak of, and medical device developers new to the field of human factors had difficulty applying its content in their domain. In addition, much of the data that yielded the design guidance in MIL-STD-1472D was obtained from young, able-bodied, male subjects, and therefore may not be applicable to much of the general population who are, and will be, the primary users of medical devices in the home (e.g., the infirm and the elderly).

In 1997, the AAMI Human Engineering Committee took on the task of revising ANSI/AAMI HE 48-1993 to make it a more useful document for medical device developers and manufacturers. One of the Committee's first decisions was to divide the revised document into two parts: a process document and handbook of human factors guidance for device developers. This decision was predicated on the fact that although the FDA had mandated that developers incorporate human factors in the design process, it did not mandate how that should be accomplished.

The purpose guiding the first part of the AAMI document revision was to provide that guidance to developers. That document now has been drafted and currently is under review. Its release in final form is expected early in 2001. This process document outlines an approach for integrating human factors into the medical device design process, and provides developers with a significant set of sources for methods and techniques (e.g., contextual inquiry, task analysis, rapid prototyping) relevant to each stage of that process.

The second part of the revision continues to pose a major challenge to the Committee. Feedback from device developers and manufacturers strongly suggested that the need with respect to design guidance was for a handbook with copious medical examples and information sufficient for it to be a "one-stop-shopping" source of human factors design information, a monumental task, and perhaps an infeasible one. The Committee, recognizing that it cannot meet this challenge alone, is developing a prospectus for such an information resource and will try to find a funding source to support its development.

One of the key issues the Committee recognized was the fact that human factors information and data, as applied to medicine, is scattered throughout the professional literature and not always easy to access. Furthermore, information and data specific to the population likely to use medical devices, including the elderly, is even more scattered and not likely to be available in the form of design guidelines readily available to the medical device design and manufacturing community. After considering the many implications and challenges posed by potential users of this resource, the Committee has begun to consider a World Wide Web (WWW)-based approach to meeting their needs.

The current plan is to implement a phased approach, which would involve updating the design guidelines from ANSI/AAMI HE 48-1993, tailoring this material to the medical device domain, and developing new material as needed. As sections of the standard are revised and approved, they would be placed on the Web site, and the material would be searchable. In addition, ancillary materials, such as case studies and checklists, would be developed or solicited and placed on the site as they become available. Finally, the process guidance provided in Part 1 would be incorporated in the site, and copious links would be provided to useful material available elsewhere on the WWW. This approach to revising the standard is novel, at least with respect to AAMI's standard development procedures, so a number of details remain to be worked out. The Committee also will need to develop sources of support to host and maintain the WWW site, a task that is currently under way.

Incorporating Human Factors Expertise in Medical Device Companies

One of the most contentious issues that arose during the development of Part 1 in the revision of ANSI/AAMI HE 48-1993 was the extent to which human factors professionals, specifically, should be represented among the personnel in medical device companies. Obviously, human factors professionals on the Committee argued strenuously that they should be an integral part of any design team, and that medical device companies should hire qualified human factors professionals. In contrast, although not surprising, human factors consultants argued the case for human factors being conducted as a consulting activity, whereas manufacturer representatives on the Committee thought that they could develop human factors expertise in-house, if they did not already have it. This turned out to be a nonnegotiable issue. As a consequence, the human factors process guidance in Part 1 specifies only the process that should take place and, in general, the expertise required. It does not mandate how that expertise is to be acquired or deployed within a company.

IMPROVING MEDICAL DEVICE DESIGN

Incorporating human factors into the medical device design process will go a long way toward improving the final products for many users. Medical device developers, first and foremost, need to understand the capabilities and limitations of their user population, as well as the context of use, including the situations and constraints posed by the home environment. (See Beith, chap. 2, this volume, for an overview of the needs assessment process as a starting point for the application of human factors principles.) As Saladrow (1996) stated:

It is no longer logical, cost effective, or efficient to take hospital-designed equipment and automatically transfer its use to other sites of care, and to different users of the equipment. This practice will not only lead to a higher cost of delivery of healthcare, but also contribute to an overall decrease of the quality of care we provide for our patients. ... Manufacturers must incorporate into their design cycle times and resources to conduct usability testing of their products to ensure adequate human factors design.

Manufacturers must perform these usability tests in conjunction with providers who are responsible for providing the care to patients in the environments in which the products and equipment will be used. Failure to do this will not allow manufacturers to produce well-designed effective products and will not permit providers and managed care organizations to achieve the most optimum delivery of care. (pp. S21 and S23)

Part 1 in the revision of ANSI/AAMI HE-48 will contain an extensive, annotated list of resources on human factors engineering tools and techniques including task analysis, contextual inquiry, simulation, heuristic analysis, workload assessment, and usability testing. Among the resources cited for usability testing, for example, are Nielsen (1993), Dumas and Redish (1999), and Rubin (1994).

Besides incorporating human factors in the medical device design process, developers should make use of the specific design guidance available to solve existing design problems that indeed are solvable. One human factors engineering consultant, who has worked for many years in the medical device industry, has provided specific design tips for some of the more common problems in medical device display design. As can be seen in the sample that follows (Wiklund, 1998), applying these tips during design would improve many medical devices, not only for the elderly, but for almost all users.

Reduce screen density.   Keep displays simple so they do not intimidate users. Secondary sources of information should be relegated to pop-up menus, for example.

Provide navigation cues and options.   Number pages, use meaningful titles on screens, and provide obvious visual cues allowing users to move forward and backward, get to the main menu, or access help screens.

Limit the number of colors.   Limit the number of colors in displays, and ensure they are consistent with medical conventions.

Simplify typography.   Specific guidance includes committing to a single font, ensuring that the most important information is the most easily read, and limiting the use of highlighting.

Use simple language.   What may be "old hat" to designers may constitute complex jargon for laypersons.

Refine and harmonize icons.   Invest in the design of quality icons that are simple and readily recognizable to users.

These tips are simple, but they have not been applied very often in the medical device domain. Many of Wiklund's (1998) ideas apply just as readily to documentation and training manual design, an area in particular need of attention as medical devices are, more and more, considered to be consumer products. Patricia Wright (1999a, 1999b, 2000), whose research in documentation design is well known, has recently published a number of articles concerning the design of health care documentation specifically, one of which targets the elderly as consumers of documentation (Wright, in press). In that article, she addresses both the sensory and cognitive needs of the elderly with respect to documentation. As previously discussed, because so many of the elderly have sensory deficits of one sort or another, it is very helpful for medical devices to use redundancy in information presentation (e.g., simultaneous visual and auditory displays of the same information). Similarly, because many older adults have difficulties remembering sequences of instructions, designers of medical devices, user manuals, and documentation can enhance their designs in many ways to help manage this problem. Among the design options available are the following:

Decrease the number of instruction steps required.

Include pictures or other visual cues that can help users map their actions to the documentation steps.

Keep sequences of instructions visually available to the user during task performance, eliminating the need for recall of steps.

Provide feedback for each user action that allows users to keep track of where they are in a sequence of instructions. (For example, given a visually displayed instructional sequence, highlight with an arrow the next step the user should perform as the prior step is completed.)

Other excellent sources of information on documentation design for older adults include Morrell and Echt (1997), Hartley (1999), and Callan (1996), the latter dealing specifically with the design of user instructions for medical devices. Device manufacturers would do well to use this information, in addition to the FDA's writing style guide, *Write it Right* (Backinger & Kingsley, 1993), in the design of device documentation that likely will be used by the elderly.

Device developers and manufacturers also need to recognize that medical documentation can benefit from usability testing, just as medical devices do. Schriver (1997), who has conducted such testing, noted that usability evaluation prompted document designers to deal with problems they otherwise would have missed or ignored, and that through the testing they gained a new perspective unavailable to them previously, despite years of experience in documentation design. The design of usability evaluations for documentation and medical devices should take into account the capabilities and limitations of the elderly population when this population will be involved in the testing. Older subjects, for example, are likely to require more time to complete usability evaluation tasks, and may require instructions minimizing the amount information that must be maintained in short-term memory to complete the tasks in the evaluation.

Also of importance is the delivery of training in the use, care, and maintenance of medical devices. Training in context (e.g., in the home) will be far more effective than training in the hospital, for reasons discussed previously. In addition, supervised training in the home environment, with a professional competent in the use of the device present, will allow better handling and management of special environmental conditions that may exist in a particular home environment.

When constructing training protocols for elderly device users, developers should keep in mind the deficits and learning difficulties frequently encountered by older adults. Cohen (1988), for example, has shown that older adults have greater difficulty than younger adults completing tasks that require inferences based on information presented. The elderly also have greater difficulty comprehending material when comprehension requires that they maintain in short-term memory information they read previously in order to interpret the new information (Light, 1992). Finally, older adults require more training time than younger adults because they learn at a slower rate, and this difference becomes more pronounced as the material to be learned becomes more complex (Gwynne, Kelly, & Callan, 1991).

CONCLUSION

In many ways, the medical field, as a whole, is playing catch-up, as compared with other industries, regarding incorporation of human factors engineering processes and principles into the design of medical devices and equipment. The situation is particularly critical now that more and more medical devices are being placed in the home and used by lay caregivers, many of whom are elderly and have significant physical, sensory, and cognitive limitations. Adoption of sound human factors practices that involve end users throughout the design process will improve medical device design and increase the safety, usability, and acceptance of medical devices.

Many of the barriers to achieving this goal are being struck down as a result of the new FDA regulations and the increasing media coverage of human error in medicine (Institute of Medicine, 1999). When the respected Institute of Medicine notes that, even by conservative estimates, more people die annually from medical errors than from workplace accidents, acquired immunodeficiency syndrome (AIDS), highway accidents, and breast cancer, the public is likely to take notice and demand action. It remains to be seen, however, whether the FDA's new regulations will be enforced. This would require, among other things, a significant investment in the training of FDA inspectors, who, until recently, were not trained with respect to human factors engineering.

A significant barrier that remains is the usable packaging of human factors design guidance specific to the user population that can serve as a resource to device developers. Much information about the capabilities, limitations, and needs of the elderly is available, but it is difficult to access or to apply directly as design guidance because it is not presented in a form usable by design engineers. With respect to some design elements and design decisions, information specific to the elderly population simply is not yet available. Hopefully, as device manufacturers begin to apply human factors methods and techniques, and to study their users and the environments of device use, the human factors knowledge base for the design of medical devices to be used by the elderly will increase.

Finally, because the patient will remain the final line of defense against medical error in a far-from-perfect system, the elderly themselves need to guard against overreliance on the health care system, including their caregivers, by asking questions, pointing out problems, and advocating for devices and documentation that meet their needs. As Lachman (1991) has shown, older people are not necessarily destined to accept a lack of control and a total reliance on others to make decisions appropriate to their health care. If training regimens are designed that allow elders to use technology effectively and succeed in self-care and caregiving activities despite any deficits or impairments they may have, their willingness to participate actively in the betterment of the health care system and its associated technologies may well increase.

REFERENCES

American National Standards Institute/Association for the Advancement of Medical Instrumentation. (1988/1993). *Human factors engineering guidelines and preferred practices for the design of medical devices* (ANSI/AAMI HE 48–1993). Washington, DC: Author.

Bach, S. (2000, February 16). *Medical mistakes and professional responsibility.* Presentation sponsored by the Center for the Study of Ethics in Society, Western Michigan University, Kalamazoo, MI.

Backinger, C. L., & Kingsley, P. A. (1993, August). *Write it right: Recommendations for developing user instruction manuals for medical devices used in home health care* (HHS Publication FDA 93–4258). Washington, DC: Food and Drug Administration, Center for Devices and Radiological Health.

Berglund, S. (1998). Systems failures, human error, and healthcare. *Medical Liability Monitor,* 1–4.

Bogner, M. S. (1993). Medical devices: A new frontier for human factors. *CSERIAC Gateway, 4*(1), 12–14.

Boodman, S. G. (1999, November 30). Handle with care: Increasingly, parents are expected to treat babies with serious medical problems at home. *Washington Post,* Z10.

Bridger, R. S., & Poluta, M. A. (1998, May/June). Ergonomics: Introducing the human factor into the clinical setting. *Journal of Clinical Engineering, 23*(3), 180–188.

Brock, D. B., & Brody, J. A. (1985). Statistical and epidemiologic characteristics. In R. Andrews, E. Bierman, & W. Hazzard (Eds.), *Principles of geriatric medicine.* McGraw-Hill.

Caplan, L. J., & Schooler, C. (1990). The effects of analogical training models and age on problem solving in a new domain. *Experimental Aging Research, 16,* 151–154.

Callan, J. R., Gwynne, J. W., & Sawyer, C. R. (1995). The role of labeling in the compliant use of medical devices. *Medical Device & Diagnostic Industry, 17*(1), 202–210.

Casali, S. P., & Chase, J. (1993). The effects of physical attributes of computer interface design on novice and experienced performance of users with physical disabilities. In *Proceedings of the 37th Annual Meeting of the Human Factors and Ergonomics Society* (pp. 849–853). Santa Monica, CA: Human Factors and Ergonomics Society.

Casarett, D., & Helms, C. (1999). Systems errors versus physicians' errors: Finding the balance in medical education. *Academic Medicine, 74*(1), 19–22.

Charness, N., Bosman, E. A., & Elliot, R. G. (1995, August). *Senior-friendly input devices: Is the pen mightier than the mouse?* Paper presented at the 103rd Annual Convention of the American Psychological Association, New York, NY.

Code of Federal Regulations. (2000, April 1) Title 21. *Final rule: Quality system regulation.* (pp. 141–142). Washington, DC: Food and Drug Administration.

Cohen, G. (1988). Age differences in memory for text: Production deficiency or processing limitations? In L. L. Light & D. M. Burke (Eds.), *Language, memory, and aging* (pp. 171–190). New York: Cambridge University Press.

COMSIS Corporation. (1988). *Product safety and the older consumer: What manufacturers/designers need to consider* (CPSC Publication 702). Washington, DC: Consumer Product Safety Commission.

Department of Defense. (1989, March 14). *Military standard: Human engineering design criteria for military systems, equipment, and facilities* (MIL-STD-1472D). Washington, DC: Author.

Dumas, J., & Redish, J. C. (1999). *A practical guide to usability testing* (rev. ed.). Bristol, UK: Intellect Books.

Dwyer, K. (1997, Winter). The role of human factors research in reducing medical errors: A conversation with Dr. Lucian Leape. *Forum, 17,*(5–6), 7–8.

Feinstein, A. R. (1997). System, supervision, standards, and the "epidemic" of negligent medical errors. *Archives of Internal Medicine, 157,* 1285–1289.

Gardner-Bonneau, D. J. (1997, November). The critical role of human factors in systems and devices for home health care. Paper presented at Medical Design & Device Manufacturing (MDDM), Minneapolis, MN.

Gardner-Bonneau, D. J., & Gosbee, J. W. (1997). Health care and rehabilitation. In A. D. Fisk & W. A. Rogers (Eds.), *Handbook of human factors and the older adult* (pp. 231–255). San Diego: Academic Press.

Gwynne, J. W., Kelly, R. T., & Callan, J. R. (1991, March). *Home medical device design for the elderly. Phase I Report* (Phase I SBIR Grant No. 1R43AG08896–01). Bethesda, MD: National Institute on Aging.

Hartley, J. (1999). What does it say? Text design, medical information, and older readers. In D. C. Park, R. W. Morrell, & K. Shifren (Eds.), *Processing of medical information in aging patients: Cognitive and human factors perspectives* (pp. 233–247). Mahwah, NJ: Lawrence Erlbaum Associates.

Hasher, L., & Zacks, R. T. (1988). Working memory, comprehension, and aging: A review and a new view. In G. H. Bower (Ed.), *The psychology of learning and motivation* (Vol. 22, pp. 193–225). San Diego: Academic Press.

Institute of Medicine. (1999). *To err is human: Building a safer health system.* Washington, DC: National Academy Press.

Kane, M. J., Hasher, L., Stoltzfus, E. R., Zacks, R. T., & Connelly, S. L. (1994). Inhibitory attentional mechanisms and aging. *Psychology and Aging, 9,* 103–112.

Klatzky, R. L., Kober, N., & Mavor, A. (Eds.) (1996). *Safe, comfortable, attractive, and easy to use: Improving the usability of home medical devices.* Report of a Workshop conducted by the Committee on Human Factors, Commission on Behavioral and Social Sciences and Education, National Research Council. Washington, DC: National Academy Press.

Kline, D. W., & Scialfa, C. T. (1997). Sensory and perceptual functioning: Basic research and human factors implications. In A. D. Fisk & W. A. Rogers (Eds.), *Handbook of human factors and the older adult* (pp. 27–54). New York: Academic Press.

Koncelik, J. (1982). *Aging and the product environment.* Florence, KY: Scientific and Academic Additions.

Lachman, M. E. (1991). Perceived control over memory aging: Developmental and intervention perspectives. *Journal of Social Issues, 47,* 159–175.

Laux, L. (1994). Visual interpretation of blood glucose test strips. *The Diabetics Educator, 20*(1), 41–44.

Leape, L. L. (1994). Error in medicine. *JAMA, 272*(23), 1851–1857.

Light, L. L. (1992). The organization of memory in old age. In F. I. M. Craik & T. A. Salthouse (Eds.), *The handbook of aging and cognition* (pp. 111–166). Hillsdale, NJ: Lawrence Erlbaum Associates.

McDowd, J., Vercruyssen, M., & Berrin, J. E. (1991). Aging, divided attention, and dual-task performance. In D. L. Damos (Ed.), *Multiple-task performance* (pp. 387–414). London: Taylor & Francis.

Morrell, R. W., & Echt, K. V. (1997). Instructional design for older computer users: The influence of cognitive factors. In W. A. Rogers, A. D. Fisk, & N. Walker (Eds.), *Aging and skilled performance: Advances in theory and application* (pp. 241–265). Hillsdale, NJ: Lawrence Erlbaum Associates.

Morrow, D., & Leirer, V. O. (1997). Aging, pilot performance, and expertise. In A. D. Fisk & W. A. Rogers (Eds.), *Handbook of human factors and the older adult* (pp. 199–230). San Diego: Academic Press.

Mosenkis, R. (1994). Human factors in design. In C. W. D. Van Gruting (Ed.), *Medical devices: International perspectives on health and safety* (pp. 41–51). Amsterdam: Elsevier.

Nielsen, J. (1993). *Usability engineering.* San Diego: Academic Press.

Rubin, J. (1994). *Handbook of usability testing: How to plan.* New York: Wiley.

Saladrow, J. (1996, May/June). Continuum of care and human factors design issues. *Medical Device & Diagnostic Industry, 19*(3S), S20–S24.

Salthouse, T. A., & Babcock, R. L. (1991). Decomposing adult age differences in working memory. *Developmental Psychology, 27,* 763–776.

Schriver, K. A. (1997). *Dynamics in document design.* Chichester, UK: Wiley.

Small, A. (1987). Design for older people. In G. Salvendy (Ed.), *Handbook of human factors* (pp. 499–500). New York: John Wiley.

Tun, P. A., & Wingfield, A. (1994). Speech recall under heavy load conditions: Age, predictability and limits on dual task interference. *Aging and Cognition, 1,* 29–44.

U. S. Food and Drug Administration. (1998a, February 14). *Current good manufacturing practice (CGMP): Final rule, quality systems regulation.* Washington, DC: Author.

U. S. Food and Drug Administration. (1998b). Minimizing medical product errors: A systems approach. *Food and Drug Administration,* 1–13.

White, L. R., Cartwright, W. S., Cornoni-Huntley, J., & Brock, D. B. (1986). Geriatric epidemiology. In C. Eisdorfer (Ed.), *Annual review of gerontology and geriatrics, 6* (pp. 215–311). New York: Springer.

Wiklund, M. (1998, May). Making medical device interfaces more user-friendly. *Medical Device & Diagnostic Industry,* 177–182.

Wright, P. (1999a). Writing and information design of healthcare materials. In C. Candlin & K. Hyland (Eds.), *Writing: Texts, processes, and practices* (pp. 85–98). London: Addison-Wesley Longman.

Wright, P. (1999b). Comprehension of printed instructions: Examples from health materials. In D. Wagner, R. Venezky, & B. Street (Eds.), *Literacy: An international handbook* (pp. 192–198). Boulder, CO: Westview Press.

Wright, P. (2000). Supportive documentation for older people. In P. H. Westendorp, C. J. M. Jansen, & R. Punselie (Eds.), *Interface design and documentation design* (pp. 81–100). Amsterdam/Atlanta: Rodopi.

Zacks, R., & Hasher, L. (1997). Cognitive gerontology and attentional inhibition: A reply to Burke and McDowd. *Journal of Gerontology: Psychological Sciences, 52B*(6), P274–P283.

PART V
COMPUTERS AND HEALTH CARE

13

Computer Interface Issues for Health Self-Care: Cognitive and Perceptual Constraints

Neil Charness
Patricia Holley
Florida State University

The rising cost of health care has provided a strong incentive within the health care system to move users from institution-based to home-based delivery of medical treatment. Therefore, tasks that previously required the intervention of trained health care providers increasingly fall on patients and their families. Technological innovation is seen as a way for allowing patients to take a larger, more direct role in maintaining their health. Unfortunately, most medical devices are not designed for patient self-care, but for use by trained medical personnel. This chapter discusses aging issues relevant to interface design, beginning with a review of broad trends in health care in the United States that provide the justification for considering human factors issues in patient self-care.

As health care devices become more sophisticated, they make increasing demands on the waning cognitive, perceptual, and psychomotor capabilities of older adult users. By taking a human factors approach, researchers can begin to address how to design an interface sensitive to the needs of an aging population. Human factors issues arise in two areas: design of the physical interface for the device and design of instruction for using and maintaining the device. This chapter reviews age-related changes in cognitive and perceptual capabilities and their likely impact on the design of interfaces, concentrating on programmable devices. Also considered is the role of cognitive and perceptual abilities as well as attitudes toward and experience with technology as risk factors for the successful use of medical technology.

One motivation for concern about health care device use by older adults and their caregivers is the possible increased risk for error. As the Hippocratic oath clearly states: "First, do no harm!" Unsuccessful operation of medical devices and procedures by health care professionals has caught the public's attention in several recently released reports, such as the 1999 Institute of Medicine report, To Err Is Human: Building a Safer Health System, which concluded that hospital care errors may result in more than 100,000 unnecessary deaths per year. A *New York Times* article, Death by Prescription: The Boom in Medications Brings Rise in Fatal Risks by Stolberg (June 3, 1999), outlines the problems encountered in filling a prescription through a pharmacy, even in hospital settings. One possible way to prevent these errors is to redesign prescription filling through technology. Stolberg outlined some exemplary computer-based entry systems that check the prescription for spelling errors, overdoses, and potential drug interactions (based on checking the patient's medical record), thereafter sending the order electronically to the pharmacy. Such systems significantly reduced physician prescription error rates.

However, most computer systems, at least those that many of us use, are neither 100% reliable nor particularly user-friendly. They also introduce the possibility of error not only through user input, but also through programmer oversights in the designing of software, as in the case of the Therac-25 radiation therapy accidents (Leveson & Turner, 1993). Therefore, the introduction of computer technology into health care brings with it a complicated set of human factors issues.

SOCIETAL TRENDS

It is well known that the U.S. population is aging. The graph in Fig. 13.1 is based on data from the Census Bureau Web site: http://www.census.gov.

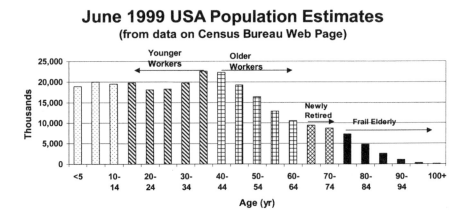

FIG. 13.1.   U.S. population profile in 1999. (Source for data was http://www.census.gov/ from the Resident Population Estimates of the United States by Age and Sex: April 1, 1990 to June 1, 1999.

The so-called "baby-boom" cohorts are those currently 35 to 39 years through 50 to 54 years of age. As outlined elsewhere (Charness, 1998), it is convenient to break the population into bands such as younger and older workers, newly retired adults, and the so-called frail elderly. The latter group is most at risk for perceptual, cognitive, and psychomotor impairments that can interfere with successful functioning using medical devices. The various cohorts shown in Fig. 13.1 will most likely be faced with different challenges to be surmounted. Aging workers probably will need to be able to use medical devices successfully within the practice of their health care professions. Newly retired adults likely will be dealing primarily with managing minor chronic health care conditions, most probably pill-taking regimens. The frail elderly adult population probably will have to deal with managing more serious acute and chronic disease processes.

WHO ARE THE PROVIDERS?

Public and private payers purchase health care from a variety of providers. Knowing who the providers are and where the largest expenditures occur suggests where redesign of interfaces may have the most impact in the health care industry. Figure 13.2 shows who pays and who is paid for health care, as exemplified by 1997 expenditures.

Physicians and hospitals are the major recipients of payments, although prescription drug costs have accelerated such that prescription drug costs now exceed nursing home costs. Indeed, there is justification for concern over the accuracy of filling prescriptions, given their cost, which probably reflects higher numbers of more costly drugs and an increase in the number of drugs prescribed. (http://www.hcfa.gov/stats/indicatr/tables/t08.htm, accessed 9/20/00.)

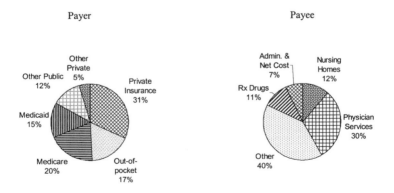

FIG. 13.2.   U.S. percentage expenditures on health care: Payers and payees.

Figure 13.3 outlines trends in expenditures. The top indicates private, public, and total spending. The bottom provides information about spending in dollars by sector.

Clearly, if the aim is to reduce societal costs of health care, then hospital facilities and physicians will need to take the biggest reductions. Health maintenance organizations (HMOs) and the federal government as administrator of Medicare and Medicaid are moving in this direction by bargaining for cheaper rates from hospitals and physicians on behalf of their members, as well as by restricting access to those resources. What this means is increased reliance on less highly trained and paid professionals: for example, using nurse practitioners in place of physicians, registered nurse assistants in place of registered nurses, and less skilled lab technicians in place of more skilled ones by automating test procedures. Furthermore, health care recipients and their caregivers are being asked increasingly to provide more of their own services. In short, there is a slowing in the previous trend of transferring costs from the private to the public sector as seen in the Fig. 13.4.

The preceding figures provide the macrotrends in health care expenditures. In less than 40 years, there has been a conversion from a health care system funded al-

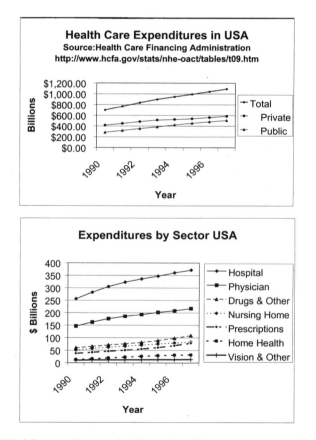

FIG. 13.3.   U.S. dollar expenditures on health care by public and private sources (top) and by sector (bottom).

Percent Health Expenditures USA

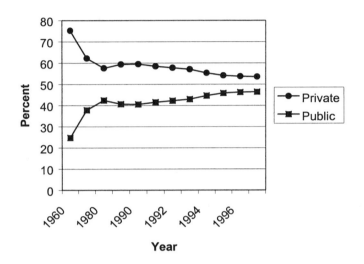

FIG. 13.4.   Percentage of U.S. health care expenditures by public and private sectors over time.

most totally by private payers to one evenly split between private and public payers. The recent conservative political trend in the United States to diminish government involvement in favor of private sector control inevitably will bring in more do-it-yourself health care. There is a broad trend in society to automate services, driven in part by widespread availability of computers and distance communication through telephone devices. Currently, automated phone systems force consumers of all ages to connect themselves to appropriate information sources rather than knowledgeable human service providers. Businesses invite consumers to carry out transactions that paid employees used to do for them (e.g., prescription refill requests, retrieval test results).

These are not entirely new trends. Sears introduced remote (catalog) sales in the 1890s, with literacy as the basic requirement for the user. Now, however, the cognitive requirements for consumers of health care probably are increasing from those of a century ago. Home health kits are replacing laboratory-administered tests for everything from pregnancy to human immunodeficiency virus (HIV) testing to blood glucose monitoring.

WHAT ARE HEALTH CARE CONSUMERS BEING ASKED TO DO?

It is useful to start with a definition of medical devices. Given that federal or state governments usually regulate such devices, one such case is examined. The German government defined medical devices via (Law on Medical Devices, 1995) (http://www.bfarm.de/gb_ver/medicaldev/, accessed 12/15/99):

The term medical devices means any instrument, apparatus, appliance, substances, and preparations from substances intended by the manufacturer to be used for human beings. Their function can be (a) diagnosis, prevention, monitoring, treatment, or alleviation of disease; (b) diagnosis, monitoring, treatment, alleviation of or compensation for an injury or handicap; (c) investigation, replacement, or modification of the anatomy or of a physiological process; or (d) control of conception.

Devices usually are broken into classes related to safety and effectiveness. In the United States, the Food and Drug Administration (FDA) is responsible for overseeing medical devices and uses a Class I, Class II, Class III system. Class I devices have the least regulatory control, and the potential for harm to the user is minimal. Examples of Class I devices include elastic bandages, examination gloves, and hand-held surgical instruments. Class II devices require more than general controls to ensure safety and effectiveness. They also are subject to special controls that may include special labeling requirements, guidance documents, mandatory performance standards, and postmarket surveillance. Examples of Class II devices include powered wheelchairs, infusion pumps, and surgical drapes. Class III devices are defined as those that support or sustain human life, assume substantial importance in preventing impairment of human health, or prevent a potential, unreasonable risk of illness or injury. An example would be an implantable pacemaker pulse generator (http://www.fda.gov/cdrh/dsma/dsmaclas, accessed 9/21/00). Devices are grouped under separate panels (such as, anesthesiology, ophthalmology, or radiology (http://www.fda.gov/cdrh/devadvice/3131.html#contents; accessed 12/15/1999).

This chapter focuses primarily on devices that can be used by patients or caregivers at home because these are most likely to be affected by patients' skill level or impairment. Examples of complex devices are blood pressure monitoring devices, intravenous perfusion pumps, and blood glucometers. Less complex devices are assistive and prosthetic in-home medical devices such as hearing aids. Simpler low-technology examples of medical devices are diagnostic kits (e.g., fecal blood detectors and pregnancy tests).

With the push toward self-care, patients now have to assume the role of skilled medical personnel when interacting with medical devices. Given the recent report by the Institute of Medicine on iatrogenic events (health care provider–induced injuries) in health care facilities, this may not be such a bad idea. In many cases, the equipment being used has a display that provides information about state variables (e.g., electronic thermometer), and more than likely is a programmable device that permits interaction, usually through button presses (e.g., computer-based intravenous perfusion pump). Furthermore, most medical care procedures involve multiple steps requiring careful monitoring by the patient or caregiver.

Greater burdens are placed on communication processes with distant health professionals. Today's patient may have to use a voice mail system to get HMO approval to make appointments with specialist physicians: "Enter your HMO number followed by the pound sign." Tomorrow's patients may need the ability to check a Web page for such information, and probably in the near future to use Internet communication channels to ensure that telemedicine devices are properly engaged for a videoconference session with a physician assistant or nurse practitioner. In all probability, interacting with programmable devices (primarily computers) will be a necessary task for older adults. Currently, many physician groups maintain a large corps of office workers who spend much of their time answering phone calls from patients requesting such information. One way to cut administrative costs is to automate the dispensing of such information.

WHAT COGNITIVE, PERCEPTUAL, AND PSYCHOMOTOR LIMITATIONS DO OLDER ADULTS FACE?

The demographics of aging apply equally well to aging health care professionals and aging health care recipients along with their informal caregivers. This section focuses on changes in capabilities associated with advancing age that are likely to affect the usability of programmable devices.

Vision

Older adults experience several negative age-related changes in visual functioning that should be taken into account by designers. Some issues involving the aging visual system include increased incidence of eyestrain in older adults and changes in the distance for optimal accommodation (Fozard, 1990; Hedman & Briem, 1984), both of which can affect interactions between the patient and the device if they not considered in the interface design. Similarly, the findings that older adults take longer to process visual information presented on a display screen (Camisa & Schmidt, 1984) and that they experience increased sensitivity to screen flicker (Casson, Johnson, & Nelson-Quigg, 1993; Fozard, Wolf, Bell, McFarland, & Podolsky, 1977) have important implications for interface design. Older adults also appear to experience increased visual persistence. They tend to have the sensation of vision for a longer period after a stimulus has been removed (Hawthorn, 2000; McFarland, Warren, & Karis, 1958). Something as simple as increasing the amount of time between presentations of visual information could improve usability for older adults.

Presbyopia refers to normative adverse changes in the visual system with age. The most prevalent feature is inability to focus the eyes on fine visual details at near distances. One cure for the condition is prescription of corrective lenses to bring near-vision objects into focus. What that means for much of the population wearing corrective lenses for far vision (i.e., people who are myopic in their youth) is the adoption of multilens systems such as bifocal or progressive lenses in spectacles or multiband contact lenses. Even with such correction in place, there still are difficulties. First, some people have difficulty adapting to multiple lens systems. Second, even with progressive lenses, which provide a gradient of correction, the size of the effective visual field is relatively small. The visual field also becomes functionally smaller with increasing age (Ball, Owsley, & Beard, 1990). This can produce some difficulties in working with large documents or computer displays, which usually are overcome by making head rather than eye movements to focus on items of interest.

A second approach to making fine visual detail more easily accessible is to increase the size of the display or the items in the display (e.g., increasing font size for print). In the case of medical devices with small displays (e.g., electronic thermometers), changing display size may involve considerable redesign cost, often because the market for such devices is fairly small to begin with. (Still, the electronic thermometer is a major step up from the old mercury thermometers for ease of reading temperature.) Other approaches, such as the use of magnifying lens systems, are possible, but bring other risks such as occupying a limb needed for operating the device. There may not be an unimpaired limb available because of age-related increases in arthritis.

Another factor that can limit vision is ambient luminance. As Charness and Dijkstra (1999) showed, home environments usually have luminance levels far below recommended levels for reading, particularly for reading by older adults. The older eye admits about one third of the light admitted by the eye of a 20-year-old, so low luminance can be particularly impairing for older adults. Fortunately, this problem is solved easily by inexpensive retrofitting such as adding lamps.

The more likely approach to resolving vision problems is to display information on a large screen. Hospitals already hook up multiple medical devices to multidisplay devices. Homes are experiencing similar technology for display of videocassette recorder (VCR) and television settings. So-called "smart homes" with built-in networking will make this feasible in the longer term, and inexpensive bridging technologies such as infrared and high-frequency radio broadcast techniques can be used now.

Audition

Older adults typically experience a decline in sensitivity from approximately their preteen years as measured by pure tone thresholds. In addition, they begin to lose the ability to separate signal from the signal-plus-noise environment. Disorders such as *tinnitus* (internal noise perceived as a constant background sound such as buzzing) that increase with age (approximately 6% to 7% of adults older than 75 years of age report having tinnitus; Benson & Marano, 1998) can dramatically degrade hearing. In addition to these problems, older adults experience difficulties discriminating speech sounds, even with the help of contextual information (Stine-Morrow, Miller, & Nevin, 1999).

Even with declines in hearing sensitivity from young adulthood for pure tone thresholds, speech discrimination (e.g., ability to identify monosyllabic words) tends to improve into the 20s, remains fairly constant through the 40s, and then shows a fairly sharp decline after the 50s, particularly for men (Jerger, 1973). In fact, the speech discrimination performance of older adults was consistently worse in noisy situations, even when matched on hearing threshold sensitivity for speech and pure-tone stimuli with that of younger adults (Dubno, Dirks, & Morgan, 1984). Dubno et al. (1984) found that an age factor contributed to an overall difference between younger and older adults in ability to recognize speech in noise, regardless of hearing loss, for conversational speech levels ranging from soft (56 dB) to loud (88 dB), and reported similar findings for speech recognition in the absence of noise. Typical nonoptimal conditions encountered in daily life include rapid speech, echoed speech (public announcements), and the ever-present irrelevant speech of multiperson conversations.

A loss in hearing acuity may affect the ability to focus attention effectively. According to Murphy, McDowd, and Wilcox (1999), the loss may affect the ability to orient and attend to auditory information in the face of competing sound. Research from the Berlin BASE project identified poor hearing acuity as a risk factor for cognitive impairment (Lindenberger & Baltes, 1994). Poor hearing acuity also may impair social functioning (Wahl & Tesch-Römer, 2001).

Crocker (1997) suggested that noise also can adversely affect communication that relies on visual channels, specifically, two effects of noise on reading performance. First, loud nonspeech noise (68 dB and above) disrupts reading comprehension. The presence of background irrelevant speech disrupts comprehension and memory processes even at relatively low levels (40–76 dB). This effect probably results from competition for attentional resources by speech-processing mechanisms and by the comprehension processes involved in reading. (See Qualls, Harris, & Rogers, chap. 4, this volume, for a more detailed review of cognitive-linguistic factors in comprehension.) It is possible that these effects might be even larger for older adults because these results were based on an adult (but not older adult) population, particularly given the greater distractibility or inability to inhibit irrelevant information of older adults found in some conditions (Carlson, Hasher, Connelly, & Zacks, 1995). One implication of this work is that multimedia communication systems could be more difficult for older adults, particularly in the one-way situation of transmitting to multiple participants when there is asynchrony between spoken and visually presented information. (See Whitten, chap. 7, and Czaja, chap. 9, this volume for a review of factors constraining effective use of telecommunication technology.)

Psychomotor Control

The primary changes in psychomotor function with increased age are a slowing in response speed (Salthouse, 1996) and declines in motor coordination and dexterity (Avolio & Waldman, 1994), the latter evident even during the working years (20s through 60s). However, more specific deficits have shown up that have a direct impact on computer technology. Walker, Philbin, and Fisk (1997) have shown specific age disadvantages for older as compared with younger adults in controlling a mouse input device, although modification of movement parameters with existing mice (e.g., slowing them down) and better design of the software supporting mouse movement can help. Smith, Sharit, and Czaja (1999) replicated the finding of difficulty for older adults with a mouse input device. Charness, Kelley, Bosman, and Mottram (2001) showed that a subset of older novice users being trained to use word processing software had considerable difficulty using a mouse. However, in the latter study, it was found that a group of older experienced computer users, despite their inexperience as mouse users, showed no such deficit. Another common problem with increasing age, affecting 40% to 60% of the general population older than 65 years is arthritis (Verbrugge, Lepkowski, & Konkol, 1991), which when experienced in the hands and wrists can seriously impair the use of pointing devices and keyboards.

Factors such as arthritis and loss of general hand and finger dexterity can play a role in impaired control of the usual input devices for computers (e.g., keyboard and mouse). The latter is probably more important than the former for producing impairment, although irritation associated with control aspects of a device (e.g., age-related slowing, loss of dexterity) may be sufficient to discourage use. Whether surfing the Web to find medical information today or videoconferencing with medical personnel in the future, older adults who have difficulty operating input devices such as a mouse are going to be seriously disadvantaged in self-care situations.

Current generations of aging workers, and hence future generations of older adults, may include subsets that have been injured at work through repeated trauma disorders:

> The most common repetitive task associated with repeated trauma disorders, according to Bureau of Labor Statistics survey data, is placing, grasping, or moving objects other than tools (for example, scanning groceries at the checkout counter). Other work activities, such as typing or key entry, and repetitive use of tools also produce large numbers of repeated trauma disorder cases. (Drudi, 1997).

These injuries are precisely the type that impair control of computer input devices such as a keyboard, mouse, or trackball.

One approach to preventing such injuries is to achieve better design of work environments and careful choice of appropriate input devices. However, a little-explored area of research involves training people to use both hands for controlling input devices, possibly lessening the burden on the preferred hand and avoiding injury by alternating hands. Information is needed on how quickly people can acquire skill with nonpreferred limbs as well as the opportunity cost while skill is being acquired, and particularly on how older adults fare with such training.

Another approach to human–computer interaction is to change the mode of input from one that relies on fine limb movement to one that relies on speech. Speech recognition software is becoming more common and more effective (90%+ accuracy: Furey, 2000). Such software generally requires a calibration period in which a given user trains the system to recognize his or her specific voice parameters. Recognition software usually cannot operate effectively without training by the user, except in instances wherein the vocabulary is very restricted, as found in automated telephone systems. Some readers will be familiar with the automated telephone dialogue that prompts the user thus: "For technical support, press or say 3."

Although speech recognition software can be expected to improve over time, it is doubtful that it will soon become a "walk up and use" feature for unlimited vocabulary situations. Furthermore, there are some diseases, fortunately rare, that so distort speech (Parkinson's disease, amyotrophic lateral sclerosis, muscular dystrophy) as to render speech recognition software nearly useless. Some conditions result in loss of speech (stroke, laryngectomy following cancer), rendering such software completely useless.

Cognition

A major constraint on the effectiveness of all health care technologies is learning how to use them. A useful way to conceptualize the impact of the myriad cognitive changes that occur with aging is as a loss in the reliability of functioning. Loss of reliability can be seen as leading to error in the completion of multistep procedures. As Fig. 13.5 shows, slight changes in reliability for completing a given step can cascade quickly into failure to complete the sequence. A system that has 99% reliability shows minimal degradation in performance even for very long sequences, but those that are 95% and 90% reliable on a given step show rapid degradation in performance as the number of steps increases. Hence, there is a strong rationale for the standard advice to minimize the number of steps in a given procedure (see Gardner-Bonneau, chap. 12, this volume.)

Research on basic memory functioning shows characteristic declines in reliability with increased adult age. However, research suggests that some memory systems are less impaired than others (e.g., semantic versus episodic memory; see chap. 3, this volume). Copious research suggests that older adults, and in some cases even middle-age adults still in the work force, do not learn as effectively as younger adults (Salthouse, 1991), that there is a general cognitive slowing that occurs with aging (Salthouse, 1996) as well as a decrease in psychomotor speed (Spirduso & MacRae, 1990). Empirical support also exists for the idea that older adults have limited processing resources (Craik & Byrd, 1982), and that older adults fail to inhibit task-irrelevant information (Hasher, Stoltzfus, Zacks, & Rypma, 1991).

General slowing and diminished efficiency in learning is evidenced in basic tasks such as a visual and memory search (Fisk & Rogers, 1991), and in complex cognitive tasks such as learning to use computer systems (Kelley & Charness, 1995). According to Salthouse (1996), a reduction in speed of information processing affects cognitive functioning by reducing the amount of time during which relevant opera-

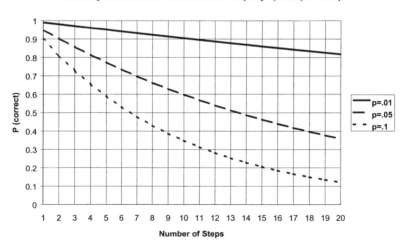

FIG. 13.5.   Success of a multistep procedure as a function of the failure probability for each step, assuming independence of failure rate by step.

tions can be executed successfully and the availability of products from earlier processes for later processing. Even in self-paced learning tasks, such as acquiring skill at word processing, older adults appear to proceed through tutorials approximately 80% slower than younger adults (Charness et al., 2001), although this is mediated somewhat by prior experience.

Similarly, when learning to use features of automated teller machines (ATMs), older adults do not acquire as much skill as younger adults following the same training (Mead & Fisk, 1998). Training procedures probably need to be tailored to older adults' capabilities and knowledge bases. More complex tasks show greater age-related differences in cognitive performance (Cerella, Poon, & Williams, 1980). However, as long as the training is constructed carefully and the procedures for use are not too complex, it is possible to show equivalent success for younger and older adults, as in the case of simulated medical laser operation (Freudenthal, 1998). Providing a mental model of device operation helped younger and older adults equally in the latter study, although this is not always the case (Caplan & Schooler, 1990). Similarly, providing environmental support in the form of external cues also may be of help to older adults (Park, Puglisi, Smith, & Dudley, 1987). Other training issues have been discussed in recent reviews (Morrell & Echt, 1997). However, there has been little effort to develop techniques for assessing cost effectiveness for training, and in the spirit of promoting this enterprise, one example is offered.

A Method for Assessing Training Effectiveness

Manufacturers of medical devices usually need to consider trade-offs between ease of use and device expense. For instance, should they manufacture an inexpensive glucometer that requires extensive instruction for successful use or a more expensive one that automates the procedure in such a way that minimal training is needed. Consumers of medical devices may make purchasing decisions based on trading off training time or money.

One approach to assessing cost effectiveness in training is to make use of learning curve data coupled with some standard economic statistics. An example is given in Fig. 13.6 below for the case of novices learning how to use a computer mouse. Data are taken from Charness, Bosman, and Elliott (1995), who investigated learning rates for different device types, such as mouse and lightpen.

Learning curve data can be used to make predictions about how many trials would be needed to enable one device to reach the performance of a superior device, or one age group to reach the performance of another. For example, if learning with the mouse is considered for the older sample, the function displayed in Fig. 13.6 is the result.

Hence, it can be predicted that it would take 64 trial blocks of 12 trials each to reach the initial young performance of 1,357 ms. Given that a block of trials took older adults approximately 5 min including breaks, it can be predicted that approximately 6 h of training time will be required for older adults to equal young worker initial performance. Taking the average industrial wage of $13.41 per hour in November, 1999, for private sector workers (Real Earnings, 1999), this means that a training investment of about $80 equates old with initial younger worker perfor-

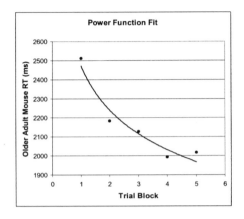

$$\text{Mouse Time} = 2480*\text{Block}^{-.145}, R^2=.95$$

Re-arranging to solve for Blocks:

$$\text{Blocks}=10^{((\text{LOG(Time)}-\text{LOG(2480))}/-0.145)}$$

FIG. 13.6 . Power function fit for older novice adults when learning to use a computer mouse.

mance. (If a replacement worker must be hired in the interim, the investment necessary might double.) Such figures, even if approximate, can help to provide useful guidelines to employers about device selection. More importantly, the training time required to bring an older worker with a mouse (that costs $15) to the same initial level as an older worker with a lightpen (that costs approximately $300) can be assessed. Substituting 1,228 ms into the equation in the right panel of Fig. 13.6 predicts the need for 127 trial blocks, approximately 12 h, or about $160 in training cost. Because of the extended training time required, a temporary worker would definitely need to be hired by the company in question, raising the cost to about $320, so it can be concluded that the employer should invest in the lightpen rather than attempt training to asymptote with a mouse. Similar analyses can be conducted to assess whether a user should purchase a given device.

However, work is definitely needed on determining the acceptability of a given medical device from the perspective of immediate usability as well as trainability. One important issue with technology introduction is trust in the device (Kantowitz, Hanowski, & Kantowitz, 1997), often a function of its perceived and actual accuracy (Dingus et al., 1997a; Dingus et al., 1997b). Probably, as with human impression formation, decisions about utility occur within the first few minutes of introduction to the interface. Still, having learning data could be useful in preventing unwarranted assumptions. It would be very useful for warning a user that it will take a given amount of time before reasonable results can be expected, such as in the case of speech recognition software that needs a calibration and training period.

CONCLUSIONS

Some evidence can be seen that health care consumers are entering an era of increasing reliance on self-care. To ensure success for older adults in this endeavor, there is a need for better information about their capabilities, with an eye toward better design of devices to fit with waning cognitive and psychomotor abilities. There also is a need to provide better instruction (minimizing steps and maximizing comprehension) for using and maintaining health care devices. The authors also suspect that if manufacturers can provide realistic estimates for the time necessary to become effective with a given device, older adults may be more willing to adopt and use health care devices.

ACKNOWLEDGMENTS

Preparation of this chapter was supported by NIA Grant P01 AG17211 for research project CREATE, an Edward R. Roybal Applied Gerontology Center.

REFERENCES

Avolio, B. J., & Waldman, D. A. (1994). Variations in cognitive, perceptual, and psychomotor abilities across the working life span: Examining the effects of race, sex, experience, education, and occupational type. *Psychology and Aging, 9,* 430–442.

Ball, K., Owsley, C., & Beard, B. (1990). Clinical visual perimetry underestimates peripheral field problems in older adults. *Clinical Vision Sciences, 5,* 113–125.

Benson, V., & Marano, M. A. (1998). *Current estimates from the National Health Interview Survey, 1995.* National Center for Health Statistics. Vital Health Stat 10 (199), pp. 79–80. Available: http://www.cdc.gov/nchswww/datawh/statab/pubd/ce95t58.htm. Accessed: 1/28/2001

Camisa, J. M., & Schmidt, M. J. (1984). Performance fatigue and stress for older VDT operators. In E. Grandjean (Ed.), *Ergonomics and health in modern offices* (pp. 270–275). London: Taylor and Francis.

Caplan, L. J., & Schooler, C. (1990). The effects of analogical training models and age on problem solving in a new domain. *Experimental Aging Research, 16,* 151–154.

Carlson, M. C., Hasher, L., Connelly, S. L., & Zacks, R. T. (1995). Aging, distraction, and the benefits of predictable location. *Psychology and Aging, 10,* 427–436.

Casson, E. J., Johnson, C. A., & Nelson-Quigg, J. M. (1993). Temporal modulation perimetry: The effects of aging and eccentricity on sensitivity in normals. *Investigative Ophthalmology and Visual Science, 34,* 3096–3102.

Cerella, J., Poon, L. W., & Williams, D. M. (1980). Age and the complexity hypothesis. In L. W. Poon (Ed.), *Aging in the nineteen-eighties: Psychological issues* (pp. 332–340). Washington, DC: American Psychological Association.

Charness, N. (1998). Ergonomics and aging: The role of interactions. In J. Graafmans, V. Taipale, & N. Charness (Eds.), *Gerontechnology: A sustainable investment in the future* (pp. 62–73). Amsterdam: IOS Press.

Charness, N., Bosman, E. A., & Elliott, R. G. (1995). *Senior-friendly input devices: Is the pen mightier than the mouse?* Paper presented at the 103rd Annual Convention of the American Psychological Association Meeting, New York.

Charness, N., & Dijkstra, K. (1999). Age, luminance, and print legibility in homes, offices, and public places. *Human Factors, 41,* 173–193.

Charness, N., Kelley, C. L., Bosman, E. A., & Mottram, M. (2001). Word processing training and retraining: Effects of adult age, experience, and interface. *Psychology and Aging, 16,* 110–127.

Craik, F. I. M., & Byrd, M. (1982). Aging and cognitive deficits: The role of attentional resources. In F. I. M. Craik & S. Trehub (Eds.), *Aging and cognitive processes* (pp. 191–211). New York: Plenum.

Crocker, M. J. (1997). Noise. In G. Salvendy (Ed.), *Handbook of human factors and ergonomics* (pp. 790–827). New York: Wiley.

Dingus, T. A., Hulse, M. C., Mollenhauer, M. A., Fleischman, R. N., McGehee, D. V., & Manakkal, N. (1997a). Effects of age, system experience, and navigation technique on driving with an Advanced Traveler Information System. *Human Factors, 39,* 177–199.

Dingus, T. A., McGehee, D. V., Manakkal, N., Jhans, S. K., Carney, C., & Hankey, J. M. (1997b). Human factors field evaluation of automotive headway maintenance/collision warning devices. *Human Factors, 39,* 216–229.

Drudi, D. (1997). BRIEF: Have disorders associated with repeated trauma stopped increasing? Available: http://stats.bls.gov/opub/cwc/1997/summer/brief4.htm. Accessed: May 19, 1999.

Dubno, J. R., Dirks, D. D., & Morgan, D. E. (1984). Effects of age and mild hearing loss on speech recognition in noise. *Journal of the Acoustical Society of America, 76,* 87–96.

Fisk, A. D., & Rogers, W. (1991). Toward an understanding of age-related memory and visual search effects. *Journal of Experimental Psychology: General, 120,* 131–149.

Fozard, J. L. (1990). Vision and hearing in aging. In J. E. Birren & K. W. Schaie (Eds.), *Handbook of mental health and aging* (3rd ed.) (pp. 150–170). San Diego: Academic Press.

Fozard, J. L., Wolf, E., Bell, B., McFarland, R. A., & Podolsky, S. (1977). Visual perception and communication. In J. E. Birren & K. W. Schaie (Eds.), *Handbook of the psychology of aging* (1st ed., pp. 497–534). New York: van Nostrand Reinhold.

Freudenthal, T. D. (1998). *Learning to use interactive devices: Age differences in the reasoning process.* ISBN 90–386–0709–1. Thesis at Eindhoven University of Technology. Eindhoven, The Netherlands.

Furey, M. J. III. (2000). Head-set free speech recognition has arrived. Available: http://www.byte.com/printableArticle?doc_id=BYT20000412S0001. Accessed: September 26, 2000.

Hasher, L., Stoltzfus, E. R., Zacks, R. T., & Rypma, B. (1991). Age and inhibition. *Journal of Experimental Psychology: Learning, Memory, and Cognition, 17,* 163–169.

Hawthorne, D. (2000). Possible implications of aging for interface designers. *Interacting with Computers, 12,* 507–528.

Health Care Financing Administration, Office of the Actuary, National Health Statistics Group: Available: http://www.hcfa.gov/stats/nhe-oact/tables/chart.htm. Accessed: November 17, 1999.

Hedman, L., & Briem, V. (1984). Focusing accuracy of VDT operators as a function of age and task. In E. Grandjean (Ed.), *Ergonomics and health in modern offices* (pp. 280–284). London: Taylor and Francis.

Jerger, J. (1973). Audiological findings in aging. *Advances in Oto-Rhino-Laryngology, 20,* 115–124.

Kantowitz, B. H., Hanowski, R. J., & Kantowitz, S. C. (1997). Driver acceptance of unreliable traffic information in familiar and unfamiliar settings. *Human Factors, 39,* 164–176.

Kelley, C. L., & Charness, N. (1995). Issues in training older adults to use computers. *Behaviour and Information Technology, 14,* 107–120.

Law on Medical Devices BFARM Federal Institute for Drugs and Medical Devices. Available: http://www.bfarm.de/gb_ver/medicaldev/. Accessed: 12/15/99.

Leveson, N., & Turner, C. S. (1993). An investigation of the Therac-25 accidents. *IEEE Computer, 25,* 18–41.

Lindenberger, U., & Baltes, P. B. (1994). Sensory functioning and intelligence in old age: A strong connection. *Psychology and Aging, 9,* 339–355.

McFarland, R. A., Warren, A. B., & Karis, C. (1958). Alteration in critical flicker frequency as a function of age and light. *Journal of Experimental Psychology, 56,* 529–538.

Mead, S., & Fisk, A. D. (1998). Measuring skill acquisition and retention with an ATM simulator: The need for age-specific training. *Human Factors, 40,* 516–523.

Morrell, R. W., & Echt, K. V. (1997). Designing written instructions for older adults: Learning to use computers. In A. D. Fisk & W. A. Rogers (Eds.), *Handbook of human factors and the older adult* (pp. 335–361). San Diego, CA: Academic Press.

Murphy, D. R., McDowd, J. M., & Wilcox, K. A. (1999). Inhibition and aging: Similarities between younger and older adults as revealed by the processing of unattended auditory information. *Psychology and Aging, 14,* 44–49.

Park, D. C., Puglisi, J. T., Smith, A. D., & Dudley, W. N. (1987). Cue utilization and encoding specificity in picture recognition by older adults. *Journal of Gerontology, 42,* 423–425.

Real Earnings. (1999). Available: http://ftp.bls.gov/pub/news.release/realer.txt. Accessed: December 29, 1999.

Salthouse, T. A. (1991). *Theoretical perspectives on cognitive aging.* Hillsdale, NJ: Lawrence Erlbaum Associates.

Salthouse, T. A. (1996). The processing-speed theory of adult age differences in cognition. *Psychological Review, 103,* 403–428.

Smith, M. W., Sharit, J., & Czaja, S. J. (1999). Aging, motor control, and the performance of computer mouse tasks. *Human Factors, 41,* 389–396.

Spirduso W. W., & MacRae, P. G. (1990). Motor performance and aging. In J. E. Birren & K. W. Schaie (Eds.), *Handbook of the psychology of aging* (3rd ed., pp. 183–220). San Diego, CA: Academic Press.

Stine-Morrow, E. A. L., Miller, L. M. S., & Nevin, J. A. (1999). The effects of context and feedback on age differences in spoken word recognition. *Journal of Gerontology, 54B,* 125–134.

Stolberg, S. G. (1999). Death by Prescription. The Boom in Medications Brings Rise in Fatal Risks. *New York Times.* Available: http://www.nytimes.com/library/national/science/060399sci-prescriptions.html. Accessed: June 3, 1999.

Verbrugge, L. M., Lepkowski, J. M., & Konkol, L. L. (1991). Levels of disability among U.S. adults with arthritis. *Journal of Gerontology, 46,* S71–S83.

USDA Food and Drug Administration. Center for Devices and Radiological Health. Available: http://www.fda.gov/cdrh/devadvice/3131.html. Accessed: December 15, 1999.

Wahl, H. W., & Tesch-Römer, C. (2001). Aging, sensory loss, and social functioning. In N. Charness, D. C. Park, & B. A. Sabel (Eds.), *Communication, technology, and aging: Opportunities and challenges for the future* (pp. 108–126). New York: Springer.

Walker, N., Philbin, D. A., & Fisk, A. D. (1997). Age-related differences in movement control: Adjusting submovement structure to optimize performance. *Journal of Gerontology: Psychological Sciences, 52B,* P40–P52.

14

Searching the World Wide Web: Can Older Adults Get What They Need?

Aideen J. Stronge
Georgia Institute of Technology

Neff Walker
UNAIDS

Wendy A. Rogers
Georgia Institute of Technology

The World Wide Web has revolutionized modern-day lives. With just the click of a few buttons, users can now access an immense amount of information, communicate worldwide, and receive numerous services. According to one survey, as of June 2000, there were 15,049,382 documented Web sites (Netcraft, 2000). Although the Web shows much promise, there is one area in which it seems to be lacking: namely, usability and specifically, usability for older adults. Two key issues involved in successfully locating information online entail locating a relevant Web site and then navigating through it for information. This chapter addresses age-related issues involved in successful information search and retrieval on the Web and how age-specific Web site design may facilitate the use of the Web by older adults.

One analogy has been drawn to compare the Web's user interface with that of the automobile (Shneiderman, 1997). Initially, automobile manufacturers each formed their own distinct design and placed their controls accordingly. Designs were revised when certain concepts proved safer than others (e.g., having the brake close to the gas pedal) and the trend emerged toward universal placement of controls such as the turn signal. Such standardization allows users to borrow someone else's car, or rent a car with little or no difficulty adjusting. Where is Web site design now with respect to the timeline of automobile generations? According to Shneiderman, Web site usability is in the Model-T stage of development. A lot of progress may have been made, but there is still a long way to go.

THE PROMISE OF THE WORLD WIDE WEB FOR OLDER USERS

The usability of any system involves the integration of three key concepts: the user population, the particular task(s) they are going to perform, and their environment. The designer then must assess any special needs this user population may require (Smith, 1997). Older adults are a user population receiving increasing attention. One reason likely reflects the number of older adults today and the predicted increases for the future. The Administration on Aging (1999) reported the following age trends:

- In the past century, the percentage of older adults has more than tripled, and life expectancy has increased by almost 30 years.

- In 1998, the population of persons older than 65 years of age numbered 34.3 million and older adults comprised 12.7% of the U.S. population (i.e., one of every eight Americans).

- By 2030, the population of older adults will increase to 70 million, and they will comprise 20% of the total U.S. population.

Evidence suggests that older adults are interested in using computers in general, and the Web in particular. For example, Czaja (1996) reported that across many studies, older adults were open-minded about the use of technology, including computers. In the past decade, older adults have shown an increased interest in using computers. In 1984, the U.S. Bureau of the Census documented that only 1% of older adults (65+ years of age) reported using a computer anywhere (Kominski & Newburger, 1999). By 1997, interest in computers had grown dramatically and 10% of older adults reported using computers.

Altogether, 7% of older adults (55+ years of age) also reported using the Internet (Kominski & Newburger, 1999). Although these numbers are impressive, older adults are still far behind younger adults in computer and Internet usage (7% vs. 25%). These age-related differences may reflect the usability of the Web because the Web does not lack information that may be of interest to older adults. Morrell, Mayhorn, and Bennett (2000) surveyed older adults and reported the following primary reasons why their sample of older adults did not use the Web:

- They did not know how the Web worked.

- They did not have access to a computer.

- They did not know what they could do on the Web.

It was not that these individuals were not interested in the Web. In fact, 38% of the nonusers in the sample were willing to learn (Morrell, personal communication).

One advantage of the Web is that it allows older adults inflicted with chronic illness the ability to access information and resources about their health. Arthritis is one example of a chronic condition that may draw older adult users to the Web in search of health-related information. The Arthritis Foundation (1999) reported that one in every six people (nearly 43 million Americans) has arthritis. This disorder is one of the most common chronic health problems and the number one cause of limitation in movement in the United States. An estimated 10% of the worldwide population is impacted by arthritis, with 50% of the people older than 60 years of age (Searle healthNet, 1999).

Older adults with arthritis may benefit by surfing the Web. One arthritis Web site offers a program called the Better Living Spa, which is designed to inform older adults how to use the Web so they may be more proactive in their own health care (Searle healthNet, 1999). This Web site consists of a Doctor's corner discussing medical management, the Juice Bar with instructions on nutrition and food preparation, the Fitness Center highlighting the importance of physical fitness, and the Locker Room complete with practical daily-living techniques. Other Web sites offer arthritis chat rooms enabling older adults to communicate with others similarly inflicted. Not only are chat rooms places to share information, but they also are the places where people are brought together in virtual support groups so they can communicate concerns, treatments, and tips with one another as well as much-needed emotional support.

Research suggests that older adult Web users do use the Web as a source for health-related information. In the Morrell et al. (2000) survey, 20% of the older users indicated that they used the Web to access health information, and nonusers listed accessing health information as something they would like to learn how to do. A similar survey, recently administered, found that accessing health information was frequently cited as an activity performed on the Web (Pak, Rogers, & Stronge, 2000).

Imagine for a moment that you are a 70-year-old just diagnosed with arthritis. Doctors and friends advise you to look for information about arthritis on the Web. Your own personal experience with the Web is limited, and quite possibly you are intimidated by the thought of trying to navigate the Web for helpful information. Where would you begin? Perhaps you could start by performing a keyword search using a Web search tool, which is an information retrieval system that assists users by pinpointing relevant Web pages on a particular topic of interest (e.g., Excite, Netscape, Google). What should you expect to find? Unfortunately, the results of your search will vary tremendously, simply as a function of the phrases and keywords you enter and the search tool you use.

To illustrate, a search of four phrases about arthritis was conducted using five well-known search tools. The four phrases chosen were "arthritis," "arthritis medication," "arthritis information," and "arthritis treatment." A glance at the results presented in Table 14.1 makes it apparent that any person, young or old, may feel overwhelmed at the sheer number of hits received (i.e., the number of Web pages re-

TABLE 14.1

Results of Search for Arthritis Keyword Phrases Conducted
on the Same Day for Five Different Web Search Tools

	Search Tool				
Keyword	Excite	Netscape	About	Google	HotBot
Arthritis	71,015	209	991	82,700	> 50,000
Arthritis medication	148,376	1	4,981	29,100	> 5,000
Arthritis information	71,015	36	118,194	114,000	> 10,000
Arthritis treatment	807,876	27	11,133	99,800	> 10,000

ported to contain the relevant phrase). Moreover, the number of hits varied greatly within each search tool for the different phrases and between the search tools. For instance, Netscape yielded 36 hits for "arthritis information," whereas About yielded 118,194 hits. Entering the keyword phrase "arthritis treatment," yields from 27 to 807,876 hits, depending on the particular search tool that was used. With this variety and number of hits, it is surprising that anyone is able to search the Web successfully for information.

The investigation of the Web's promise for older adults showed the following trends. Older adults are interested in using the Web for a variety of activities, one of which is searching for health-related information. Many older adults already use the Web, although it is not known how successfully. This point is discussed later. Older adults who do not yet use the Web express willingness to learn. With the many advantages associated with use of the Web, the question is whether the Web is designed suitably for the older user.

FINDING WHAT THEY NEED ON THE WEB: INFORMATION SEARCH AND RETRIEVAL BY OLDER ADULTS

How well can older adults find the information they are searching for on the Web? Consider what an individual must do to surf the Web successfully for the answer to a question. Table 14.2 presents an abbreviated task analysis of what the person has to do. At any point in the process, there are a multitude of outcomes and numerous opportunities for an error to be made. Clearly, information search and retrieval in an automated environment requires specialized knowledge and skills for success. Borgman (1986) identified three prerequisites for information search and retrieval on an online library catalog: conceptual knowledge, semantic knowledge, and mechanical skills. *Conceptual knowledge* involves taking an information need and putting it into a searchable query (e.g., formulating queries, Boolean searching). *Semantic knowledge* involves knowing how to use the system to implement a query (i.e., how and when to use the system's features). *Mechanical* skills consist of having the basic

TABLE 14.2

Abbreviated Task Analysis for Information Searching
on the World Wide Web

Step	Description
1. Define the information need • Select keywords and phrases • Consult a thesaurus • Check spelling of terms	The first step is to frame the question and decide exactly what to look for.
2. Find the information • Go to a known URL • Use a bookmark • Follow links from a known URL • Conduct a general search • Choose a search tool	The second step is to figure out how to find the information. There are a number of options. The choice depends on the level of experience in searching and the amount of knowledge about the topic of the search.

3. Evaluate the information
 • View sites sequentially
 • Scan/scroll results
 • Make provisional judgment
   • Read and evaluate
   • Decide to broaden search
   • Decide to narrow search
	The third step is to determine whether you have found the correct information. The outcome of this step may be to return to Step 1 and reformulate the search, or return to Step 2 to search in a different way.

4. Accept/reject the information • Determine credibility of source • Create a bookmark to allow easy return to site	Once the information has been found, the decision has to made about whether to accept the information as credible and accurate

computing skills necessary to interact with the system, and knowing the syntax required to enter search queries. Although these three prerequisites describe searching an online library catalog, similar categories of knowledge are necessary for Web searching. These all are candidates for age-related differences.

Age-related differences in such specialized knowledge and skills have been found in several studies on information search and retrieval. Even older adults who have experience with an online library system experience trouble with the system. For example, Sit (1998) found that older library users had trouble performing both keyword searches and Boolean searches. Most of their performance errors resulted from a lack of conceptual knowledge or their inability to translate their need for information into a query that was searchable. Yet these were individuals who had reported experience using the online system.

In a study of novice searchers, Mead, Sit, Rogers, Jamieson, and Rousseau (2000) found that searching an online library database was more problematic for older than for younger adults, even when the participants had been matched for general computer experience. Age differences were exacerbated for older adults who did not have any computer experience. Older adults were less efficient on simple search tasks and less successful overall on more complex search tasks. In addition, the older adults were more likely to make syntax errors (e.g., entering the query incorrectly) and field specification errors (e.g., entering the wrong word in a specific field). The older adults also were less likely to use Boolean operators (e.g., and, or) to optimize their searches, and showed lower comprehension of how Boolean operators would work to broaden or narrow a search.

A similar pattern of age-related differences was observed in a study of Web searching wherein age-related differences depended on the complexity of the task (Kubeck, Miller-Albrecht, & Murphy, 1999). Participants were given questions to answer via the Web using the Yahoo search tool (e.g., name five herbs used for medicinal purposes and the use of each). Performance was described in terms of efficiency of the search and quality of the answers. Efficiency was defined as the number of steps required to find an answer, and quality was defined as the completeness or accuracy of the answers as rated by the experimenters. For the simpler tasks, which required 6 to 9 steps to find the answer, there were minimal performance differences between the young and older adults. However, for the more difficult searches (requiring 13 to 16 steps) there were age-related performance differences. For their first difficult search, older adults were less efficient than younger adults, but found equally good answers. For their second difficult search, older adults were as efficient as younger adults, but received lower quality scores than the younger adults. Thus, performance differences emerged for more difficult tasks, and there was an apparent trade off between efficiency and quality of results.

Another aspect of Web searching involves searching within a site for particular information. Mead, Spaulding, Sit, Meyer, and Walker (1997) investigated age-related differences in this type of search by having participants find the answers to questions within a constrained, 19-page Web site. Their results parallel those reported by Mead et al. (2000) and Kubeck et al. (1999). Age-related differences were not observed for the tasks requiring few moves within the site (two or fewer), but older adults were less successful than young adults when three or more moves were required to find the answer to a question.

This section began by asking the question, "How well can older adults find the information they are searching for on the Web?" The review of the few existing studies relevant to this question suggest that there are age-related differences in the ability to find information in a search and retrieval type of task, whether searching a library database or searching the Web. The differences are reduced for simple and straightforward tasks, but increase as the tasks become more complex (e.g., requiring more steps or are ill-defined). Results also suggest that such difficulties are reduced if the older adults have general computer experience or specific searching experience. However, even older adults who are experienced with the particular system of interest are not efficient nor completely accurate in their performance, as observed by Sit (1998).

HOW CAN SYSTEMS BE IMPROVED TO BE MORE USABLE FOR OLDER ADULTS?

The Web will be "usable" by older adults when designers draw on the theoretical and practical knowledge base comprising fields of human factors and cognitive aging. There is a wealth of data about how perceptual, motor, and cognitive capabilities change (or do not change) as individuals age (see e.g., Craik & Salthouse, 2000; Fisk & Rogers, 1997; National Research Council, 2000; Park & Schwartz, 2000). Designers must consider these changes when developing systems whose primary user population will be older adults.

Perceptual Considerations

One common age-related problem is a decline in visual capabilities (for a review see Kline & Scialfa, 1997). Older adults may perceive colors differently because of a yellowing of the eye lens that usually occurs with age. This yellowing causes the lens to absorb more of the short wavelengths, so that it becomes easier for older adults to see reds and oranges, but more difficult for them to discriminate colors in the blue range. After the age of 70 years, contrast sensitivity decreases, and it becomes difficult for older adults to discriminate between light and dark (Kline, Schieber, Abusaura, & Coyne, 1983).

Readability is an important issue in Web site usability. Designers need to be sensitive to the size of font that they use. When interviewed, older adults indicated a preference for a sans serif font, such as Arial or Helvetica, and reacted strongly to text that was too small such as that set in 10-point font (Ellis & Kurniawan, 2000). Not only do older adults prefer a larger font size, but their performance also increases for paper tasks when 12- to 14-point font is used (Vanderplas &Vanderplas, 1980). Therefore, font size is an important issue in Web design.

To accommodate visual decrements, designers should develop and promote size, contrast, and color guidelines. They also should develop browsers that can adjust easily for visual limitations (see also Mead et al., 2000; Nielsen, 2000)

Motor Considerations

Older adults experience difficulties with fine motor coordination such as using a computer mouse to navigate through a site or positioning the cursor on a desired target. Research has suggested that when older adults use a mouse, they are slower and less accurate than younger adults (Walker, Millians, & Worden, 1996). Older adults' main problems with mouse movements were found to be related to age differences in perceptual feedback, strategy, and increased error when applying force to a mouse (Walker, Philbin, & Fisk, 1997). Walker et al. (1996) also observed that for the very small target item (3 pixels), older adults had difficulty hitting the target. Their accuracy rate was only 75% for this target, as compared with more than 90% for the larger targets (6, 12, or 24 pixels). These data suggest that there may be a critical minimum target size, below which older individuals will be unable to use a mouse to position a cursor accurately.

Mouse control can be especially difficult for older adults suffering from a chronic condition such as arthritis. Limiting the need for mouse movement may increase older adults' performance when browsing the Web. Mead et al. (1997) discovered that whereas younger adults would scroll down the Web pages looking for information, older adults would search for information one Web page at a time. This finding suggests that although a long Web page of information is an acceptable design for younger adults, small Web pages in which there is less scrolling would be a superior design for older adults, ultimately leading to improvements in their performance.

Cognitive Considerations

Problem Solving.   Experience or lack thereof failed to explain why older adults have difficulty with information search and retrieval tasks (Mead et al., 2000). Their poorer performance may reflect a general decline in cognitive abilities, or more specifically, problem solving. *Problem solving* has been defined as "a process of assembling an appropriate sequence of component procedures (or operators) to accomplish a goal" (Carlson & Yaure, 1990, p. 484). Carlson and Yaure explained that problem-solving is a dynamic task in which the sequence of procedures is decided during the problem-solving episode. Older adults have been shown to have more difficulty than younger adults in novel problem solving tasks (e.g., Denney 1980; Hartley & Anderson, 1983). Such age-related differences may mediate Web searching performance because assembling an information query using a Web search tool frequently requires that the user specify different levels of a query (e.g., narrow or broaden a search) while immersed in the problem-solving episode.

One potential solution to aid problem solving on the Web would be to provide older adults with tools to assist them in planning their search strategies. For example, search syntax should be explicit. A very simplistic example of this is presented in Figure 14.1, in which participants can choose the type of Boolean they wish to use. Clearly, this format would have to be tested for usability, but this illustration shows what is meant by an explicit search syntax.

Designers also should provide explicit feedback on the search to aid in problem solving. For example, users should be told explicitly what type of information was searched (e.g., site names, documents, metatags). They should be provided with information about the results of the search such as what criteria were used and how the output was ordered. It should be clear how to continue the search, for example, how to search for similar or dissimilar results (see also Nielsen, 2000).

One avenue to understanding older adults' performance on tasks involving information search and retrieval on the Web will be to isolate the searching strategies of older adults who are successful at finding information on the Web to determine how they compensate for age-related cognitive declines such as decreased working memory.

Search for Words or Phrases: Connected By:

FIG. 14.1. *Simplified syntax for searching the Web.*

Working Memory.   Older adults also experience declines in working memory, which may lead to difficulties in browsing the Web. Working memory entails temporarily holding and manipulating information while engaging in a variety of cognitive tasks (Baddeley, 1986). Age differences in working memory have been well documented (see Smith, 1996, for a review). When browsing a Web site, older adults may have difficulty both remembering what they are looking for and keeping track of where they are in a site. Environmental support can assist older adults in compensating for processing deficiencies by providing them with some form of a memory aid. For example, a tool to indicate position on a Web site might help to compensate for working memory declines.

Declines in working memory also may contribute to difficulty navigating through a Web site. Mead et al. (1997) found that older adults were significantly more likely to visit previously viewed pages than younger adults. They would return to pages they had already visited, and had trouble keeping track of the current page they were viewing. Memory cues, such as putting a label on each page with the site name and the name of the individual page users are viewing may help provide environmental support to older adults.

Attention.   Aspects of attention show age-related declines (for a review see Rogers, 2000). For example, older adults have been shown to have more difficulty inhibiting irrelevant information, which means that Web page content should be restricted to that which is most relevant to the user's goal. Older adults also have more difficulty selecting among information. Such selection can be aided by consistency of design across pages within a site. For example, Mead et al. (1997) discovered that the links and browser buttons above or on the left-hand side of the interface were clicked more frequently than those found in text, on the bottom of a Web page, or on a completely separate page. They suggested placing navigation aids in the top left-hand portion of the display and making them available to the user browsing a Web site, regardless of the page he or she is viewing. Such consistency reduces the demands on the user's selective attention.

Concept Formation.   The labels on an interface act as a communication tool allowing the system to translate a user's information needs. Soto (1999) asserted that successful navigation through an interface display (e.g., automated teller machine [ATM], online library database, the Web) involves defining a goal and locating labels on a display that are semantically related to this goal. Such a process is repeated as users start at the top levels of a menu hierarchy and work their way down through the system. Indeed, Soto (1999) showed that when younger adults were presented with a task requiring them to navigate through a menu system with labels ranging in semantic similarity to the task description, performance improved as semantic similarity of the labels more closely matched the task description.

How do labels influence the usability of the Web? For starters, Soto (1999) indicated that designers typically have used informal methods to choose interface labels. Therefore, a designer may view a label as an adequate representation of a task goal, but it may in fact be quite different from a user's expectations. From the perspective of human factors, this is not acceptable. Labels should be user tested.

There also is a lack of label consistency across interfaces, which contributes to the complexity of searching the Web. For example, depending on the particular interface with which a user is interacting, the label indicating how to save a Web page to a directory may be referred to differently on one Web site interface as compared with another (bookmark vs. favorite place).

It also is the case that many Web designers are younger adults whose labels for different concepts may be ineffective for an older population. A concept is defined as one's "mental representation of a category" (Galotti, 1999, p. 251). Although, age-related differences would not be expected for familiar concepts (see chap. 3, this volume), new concepts (e.g., categories on the Web, ATM, or online library database) may lead to age-related differences in the expectations of what a label represents. These age-related differences may emerge when older adults are less experienced with a task and do not know what labels will fit their goal. For instance, in one study, older adults experienced difficulties categorizing a group of objects based on their own selection criteria (Flicker, Ferris, Crook, & Bartus, 1986). Furthermore, in another study, older adults performed more poorly than younger adults when instructed to sort unrelated words into categories, label these categories, and then recall the categories on a surprise recall test (Basden, Basden, & Bartlett,

1993). This research indicates that there may be age-related differences in the ability to categorize objects in novel situations, suggesting that older adults may be unsuccessful in deciding what labels to follow when accomplishing a task on an interface such as the Web. Unfortunately, very little research has been done in this area. As reliance on display interfaces increases, the importance of effective labeling will become crucial, and it is imperative that the potential for age-related differences in concept formation be understood if the Web is to be made a place for people of all ages.

An additional obstacle to consider in designing for an aging population is that perceptual, motor, and cognitive declines are not occurring simultaneously within an individual (Vanderheiden, 1997). For instance, an individual may have the ideal motor control necessary to interact effectively with a mouse, but may have severe working memory declines. Vanderheiden also pointed out that older adults are a diverse population aging at their own rates. Whereas one older adult may be experiencing cognitive declines, another may be having perceptual difficulties. By following general age-specific design guidelines for Web sites, designers can account not only for individuals at different levels of decline, but also for the aging population as a whole (Mead, Lamson, & Rogers, in press).

USER-CENTERED DESIGN

There is an established process of design and development that should be followed for the development of any product. The human factors approach of user-centered design can and should be applied directly to Web development. *User-centered design* is the design of tools, interfaces, training programs, and tasks to suit the needs and capabilities of users.

> The process can be summarized as follows: It begins with organizational goal-setting, which includes identification of the target user population and tasks the organization expects users to perform. Then users' special needs, goals, and expectations should be assessed. A Web site design can then be proposed and a prototype built. The process continues with iterative usability evaluation and redesign until the organization's usability goals have been achieved. (Mead, Lamson, & Rogers, in press)

Once the goals of the Web site (or the Web search tool) have been defined and the target population identified, users can be queried about their needs and preferences for the site. Interviewing users to gauge their preferences is a good starting point in Web site design, but designers must be aware of the fact that what users say they want is not necessarily what will enable them best to use the system. Reported user preferences should always be tested empirically by observing the performance of users. To illustrate, Ellis and Kurniawan (2000) found that older adults verbally expressed a preference for a Web site with fewer pages that were longer. This would require them to scroll down Web pages to obtain the information they are seeking. Although the older adults expressed this preference, they were found to experience difficulty scrolling down a Web page for information, and could not seem to understand that there was information off-screen (see also, Mead et al., 1997). It was found that it was easier for them to click a link than to scroll down a page in terms of the motor skill required. Therefore, although older adults may have verbally expressed the notion that longer pages would suit them better, their performance pointed toward shorter pages that would decrease scrolling. It is imperative that designers recognize that when considering design issues for any user population, performance may carry more weight than preference.

Next, a prototype site can be developed based on the limitations and capabilities of the user population as described earlier (Mead et al., in press), and with appeal to existing design guidelines. There are well-defined guidelines for the development of user-friendly Web sites (e.g., Nielsen, 2000). Although the recommendations may not all be empirically based, many have been adopted by developers and thus are becoming standard practice. Moreover, human factors practitioners are beginning to test particular design changes and to report the benefits and trade-offs of various design decisions (Nielsen, 1997a). The guidelines typically are not age-specific, but they still provide a reasonable starting point for the development of a prototype.

The next critical step in the process involves usability testing. Although an appeal to the cognitive aging and human factors literatures may provide an excellent basis for the development of a site, there is no substitute for usability testing (see Nielsen, 1997b, and Rubin, 1994, for primers on usability testing). Initial usability testing should be conducted on an early prototype of the site to capture major design problems. It is important to remember that usability testing of designs is an iterative process. Data are gathered from users; the design is revised; more data are gathered from users; and so on. It is critical to select users who are representative of the target user population and to select tasks for them to do that are representative of what the system would ultimately be used to do.

The approach just outlined for the development of a Web site targeted to an older adult user population has proved to be very successful. Ellis, Jankowski, and Jasper (1998) applied the concept of participatory design to the development the Michigan Aging Services System (MASS) accessible at http://MASS.iog.wayne.edu. Ellis et al. started the process by building bridges of communication between the intended users of the site and the development team. They then developed a user model to represent the users' capabilities and limitations and held brainstorming sessions with prospective user groups to map the possibilities for their system. On the basis of these initial steps, they developed a prototype, which they used to conduct usability testing via questionnaire, discussions, and online feedback from users. On the basis of feedback, they were able to improve the design of the prototype. Because their site is operational, they are able to continue gathering input from users and making plans for design improvements. The Ellis et al. approach provides an excellent illustration of how the field of human factors contributes to the design of usable systems.

CONCLUSION

As we enter the next millennium, our dependence on the Web and similar technologies will continue to increase. The first step to improving the usability of a system for older adults is to understand the age-related differences in performance. Information search and retrieval on the Web is a complex problem-solving task. Older adults have difficulty with information search and retrieval tasks such as those that involve online library systems (Mead et al., 2000; Sit, 1998) as well as the Web (Kubeck et al., 1999; Mead et al., 1997).

The number of older adults is increasing at a rapid pace, and as scientists we must shift our attention and focus on the special needs of older adults. The field of human factors has much to offer. Unfortunately, it is apparent that systematic human factors–based approaches to the development of Web sites and Web search tools are few and far between (Flanders & Willis, 1998). The potential for improvement of the Web for use by older adults is vast. The information presented in this chapter should motivate designers to capitalize on the knowledge base that exists to enable the Web to reach its potential for improving the lives of older adults.

ACKNOWLEDGMENTS

The first and third authors were supported in part by grants from the National Institutes of Health (National Institute on Aging): Grant No. P50 AG11715 under the auspices of the Center for Aging and Cognition: Health, Education, and Technology (one of the Edward R. Roybal Centers for Research on Applied Gerontology) and Grant No. R01 AG18177. This chapter is based on a presentation given at the Conference on Human Factors Interventions for the Health Care of Older Adults, February, 2000, in Destin, FL.

REFERENCES

Administration on Aging. (1999). *Profile of Older Americans.* [On-Line]. Available: www.aoa.gov/aoa/stats/profile/default.htm. Accessed: 1/24/01.

Arthritis Foundation. (1999). *Arthritis Fact Sheet.* [Online]. Available: www.arthritis.org/about. Accessed: 1/24/01.

Baddeley, A. (1986). *Working memory.* New York: Oxford University Press.

Basden, B. H., Basden, D. R., & Bartlett, K. (1993). Memory and organization in elderly subjects. *Experimental Aging Research, 19*(1), 29–38.

Borgman, C. L. (1986). Why are online databases hard to use? Lessons learned from information-retrieval studies. *Journal of the American Society for Information Science, 37,* 387–400.

Carlson, R. A., & Yaure, R. G. (1990). Practice schedules and the use of component skills in problem solving. *Journal of Experimental Psychology: Learning, Memory, and Cognition, 16,* 484–496.

Craik, F. I. M., & Salthouse, T. A. (2000). *The handbook of aging and cognition.* Mahwah, NJ: Lawrence Erlbaum Associates.

Czaja, S. J. (1996). Aging and the acquisition of computer skills. In W. A. Rogers, A. D. Fisk, & N. Walker (Eds.), *Aging and skilled performance: Advances in theory and application* (pp. 201–220). Mahwah, NJ: Lawrence Erlbaum Associates.

Denney, N. W. (1980). Task demands and problem-solving strategies in middle-aged and older adults. *Journal of Gerontology, 35,* 559–564.

Ellis, R. D., Jankowski, T. B., & Jasper, J. E. (1998). Participatory design of an Internet-based information system for aging services professionals. *The Gerontologist, 38,* 743–748.

Ellis, D. E., & Kurniawan, S. H. (2000). Increasing the usability of online information for older users: A case study in participatory design. *International Journal of Human Computer Interaction, 12*(2), 263–276.

Fisk, A. D., & Rogers, W. A. (1997). *Handbook of human factors and the older adult.* San Diego, CA: Academic Press.

Flanders, V., & Willis, M. (1998). *Web pages that suck: Learn good design by looking at bad design.* San Francisco: Sybex.

Flicker, C., Ferris, S. H., Crook, T., & Bartus, R. T. (1986). The effects of aging and dementia on concept formation as measured on an object-sorting task. *Developmental Neuropsychology, 2,* 65–72.

Galotti, K. M. (1999). *Cognitive psychology: In and out of the laboratory* (2nd ed.). Belmont, CA: Wadsworth.

Hartley, A. A., & Anderson, J. W. (1983). Task complexity and problem-solving performance in younger and older adults. *Journal of Gerontology, 38,* 72–77.

Kline, D. W., Schieber, F., Abusaura, L. C., & Coyne, A. C. (1983). Age and the visual channels: Contrast sensitivity and response speed. *Journal of Gerontology, 38,* 211–216.

Kline, D. W., & Scialfa, C. T. (1997). Sensory and perceptual functioning: Basic research and human factors implications. In A. D. Fisk & W. A. Rogers (Eds.), *Handbook of human factors and the older adult* (pp. 335–361). San Diego, CA: Academic Press.

Kominski, R., & Newburger, E. (1999). Access denied: Changes in computer ownership and use: 1984–1997. [On-line]. Available: www.census.gov/population/socdemo/computer/confpap99.pdf.

Kubeck, J. E., Miller-Albrecht, S. A., & Murphy, M. D. (1999). Finding information on the World Wide Web: Exploring older adults' exploration. *Educational Gerontology, 25,* 167–183.

Mead, S. E., Lamson, N., & Rogers, W. A. (in press). Human factors guidelines for Web site usability: Health-oriented Web sites for older adults. In R. W. Morrell (Ed.), *Older adults, health information, and the World Wide Web.* Mahwah, NJ: Lawrence Erlbaum Associates.

Mead, S. E., Sit, R. A., Rogers, W. A., Jamieson, B. A., & Rousseau, G. K. (2000). Influences of general computer experience and age on library database search performance. *Behavior and Information Technology, 19,* 107–123.

Mead, S. E., Spaulding, V. A., Sit, R. A., Meyer, B., & Walker, N. (1997). Effects of age and training on World Wide Web navigation strategies. *Proceedings of the Human Factors and Ergonomics Society 41st Annual Meeting, Albuquerque, 1,* 152–156.

Morrell, R. W., Mayhorn, C. B., & Bennett, J. (2000). A survey of World Wide Web use in middle-aged and older adults. *Human Factors, 42,* 175–182.

National Research Council. (2000). *The aging mind: Opportunities for cognitive research.* Washington, DC: National Academy Press.

Netcraft. (2000). Netcraft. [On-line]. Available online: www.netcraft.com. Accessed: 1/24/01.

Nielsen, J. (1997a). *How users read on the Web* [On-Line]. Available: http://www.useit.com/alertbox/9710a.html. Accessed: 1/24/01.

Nielsen, J. (1997b). Usability testing. In G. Salvendy (Ed.), *Handbook of human factors and ergonomics* (pp. 1543–1568). New York: Wiley.

Nielsen, J. (2000). *Designing web usability: The practice of simplicity.* Indianapolis: New Riders Publishing.

Pak, R., Rogers, W. A., & Stronge, A. J. (2000). How would you describe the World Wide Web? Analogies of the Web from users. *Proceedings of the Human Factors and Ergonomics Society 44th Annual Meeting, San Diego,* p. 172.

Park, D. C., & Schwarz, N. (2000). *Cognitive aging: A primer.* Philadelphia: Psychology Press.

Rogers, W. A. (2000). Attention and aging. In D. C. Park & N. Schwarz (Ed.), *Cognitive aging: A primer* (pp. 57–73). Philadelphia: Psychology Press.

Rubin, J. (1994). *Handbook of usability testing: How to plan, design, and conduct effective tests.* New York: Wiley.

Searle healthNet (1999). *The arthritis community and Searle: A joint effort against a chronic illness* [On-Line]. Available: www.searlehealthnet.com/arthritis.

Shneiderman, B. (1997). Designing information-abundant sites: Issues and recommendations. *International Journal of Human-Computer Studies, 47,* 5–30.

Sit, R. A. (1998). Online library catalog search performance by older adult users. *Library and Information Science Research, 20,* 115–131.

Smith, A. D. (1996). Memory. In J. E. Birren & K. W. Schaie (Eds.), *Handbook of the psychology of aging* (4th edition, pp. 236–250). San Diego, CA: Academic Press.

Smith, P.A. (1997). Virtual hierarchies and virtual networks: Some lessons from hypermedia usability research applied to the World Wide Web. *International Journal of Human-Computer Studies, 47,* 67–95.

Soto, R. (1999). Learning and performing by exploration: Label quality measured by latent semantic analysis. *Proceedings of CHI'99 Conference, Pittsburgh,* pp. 418–425.

Vanderheiden, G. C. (1997). Design for people with functional limitations resulting from disability, aging, or circumstance. In G. Salvendy (Ed.), *Handbook of human factors and ergonomics* (2nd ed., pp. 2010–2052). New York: Wiley.

Vanderplas, J. M., & Vanderplas, J. H. (1980). Some factors affecting the legibility of printed materials for older adults. *Perceptual and Motor Skills, 50,* 923–932.

Walker, N., Millians, J., & Worden, A. (1996). Mouse accelerations and performance of older computer users. *Proceedings of the Human Factors and Ergonomics Society 40th annual meeting, Philadelphia, 1,* 151–154.

Walker, N., Philbin, D. A., & Fisk, A. D. (1997). Age-related differences in movement control: Adjusting submovement structure to optimize performance. *Journal of Gerontology: Psychological Sciences, 52B,* P40–P52.

AUTHOR INDEX

SUBJECT INDEX

Page numbers followed by f indicate figures. Page numbers followed by t indicate tables.